CLEAR & SIMPLE

BIOLOGY

CHARLES R. WERT
&
PATTY KREIKEMEIER-GAFFNEY

A MONARCH BOOK
Published by Prentice Hall Press
New York, NY 10023

A Monarch Book
Published by Prentice Hall Press
A Division of Simon & Schuster, Inc.
Gulf + Western Building
One Gulf + Western Plaza
New York, New York 10023

PRENTICE HALL PRESS is a trademark of Simon & Schuster, Inc.

Designed by Mike Kelly, San Diego, California
Produced by The Word Shop, Inc., San Diego, California
Manufactured in the United States of America

1 2 3 4 5 6 7 8 9 10

Library of Congress Catalog Card Number: 86-061084
ISBN: 0-671-54786-0

CONTENTS

To Lois Schaaf, my first biology teacher
P.K.-G.

To my parents
C.R.W.

ACKNOWLEDGMENTS

The authors would like to acknowledge the love and support of their spouses during the course of this project. Thank you, Tom for the *ADDITIONAL* cooking, cleaning, and babysitting; for relinquishing computer time; and for computer technical help while working on this project.

Thank you, Margaret, for proofreading, editorial advice, and, most of all, your patience.

THE NATURE OF THE BIOLOGICAL SCIENCES

Any society which does not have a constant input of new knowledge will stagnate and collapse.

— Aristotle, *The Republic*

Science is the pursuit of knowledge leading to an understanding of the nature of the universe. It is a way of seeking universal principles of order.

There are two fundamental forms of science. **Pure science**, or **basic science**, is the intellectual exploration of a field in terms of key theoretical concepts. The role of pure science is to further man's knowledge. The second form of science, **applied science** or **technology**, is any attempt to apply the knowledge of pure science (for the betterment of mankind, hopefully). Thus, applied science is dependent on pure science.

In the ancient past, art, philosophy, religion, and science constituted one pursuit — the pursuit of knowledge and truth. As social systems evolved, however, the approaches of these four disciplines toward the pursuit of truth diverged. The most conspicuous distinctions today are between

science — concerned with the "known" (the observable, the factual, and the predictive); and

religion — concerned with the "unknown" (the experiential, and the nonpredictive — issues of faith).

Science is distinguished from the other ways human beings seek to understand the natural world in that it is limited to propositions about the natural world that can be verified objectively. Science has two separate but interacting components: — data — the collection of objective evidence based on observation, experimentation, or a combination of the two; and the organization and interpretation of data leading to the discovery of meaningful patterns and relationships.

Great advances in science are not merely additions of new data, but perceptions of new relationships among available data; in other words, the development of new ideas. The ideas of science are categorized from "most valid" to "least valid," as

law or *principle*

theory or *model*

hypothesis

hunch or *guess.*

Lowest on the scale is the **hunch**, or **educated guess**, technically known as a **heuristic estimation**. A hunch becomes a **hypothesis** when it is stated in such a way that it is testable, even if the test cannot be performed immediately. When a hypothesis has survived numerous independent tests, it becomes a **theory** or **model**. A theory that has withstood considerable testing becomes elevated to the status of a **law** or **principle**, although it is not always identified as such. The theory of evolution, for example, is no longer considered a theory in the strict usage of the term, nor has it been for almost a hundred years. As far as scientists are concerned, it is a principle.

THE SCIENTIFIC METHOD

The process by which scientists seek to determine the causal relationships in nature through critical objective analysis is known as the **scientific method**. It is a series of steps within a process of investigation in which (a) a problem is identified; (b) relevant data are gathered — *observation*; (c) the problem is defined; (d) a hypothesis is formulated; (e) the hypothesis is tested — *experimentation*; (f) the results are subjected to *analysis* and *interpretation*; and, (g) one or more generalizations are formulated based on the analysis and interpretation of data — *conclusions*.

Observation

Once a problem has been identified, the phenomenon is carefully and systematically observed. The problem is then described utilizing the relevant data collected. *Note:* Observations ultimately depend on sensory data of the observer, and are therefore inherently prone to error. Technological advances may make observations more accurate, and thus add new sensory data for the solution of a problem.

Defining the Problem and Formulating the Hypothesis

The next step requires asking the proper question(s) concerning the observations in order to define the problem. Once defined, a hypothesis can be formulated.

A **hypothesis** is a tentative, *testable* explanation for observed phenomena. In the past the formation of testable hypotheses relied heavily on the scientist's creativity and imagination. **Induction,** or **inductive logic**, is a pattern of thought important in hypothesis formation. Inductive logic begins with observations and leads to hypotheses, *proceeding from the specific to the general* — **inductive generalization**. For example, if every green apple an individual tasted were sour, the person might suspect that all green apples are sour. The suspicion would be a hypothesis based on inductive generalization.

(With the advancement of computer sciences and multivariate statistics, a phenomenon can be measured and the data analyzed statistically, which takes much of the work of inductive generalization off the shoulders of the scientist. However, the scientist is still responsible for interpreting the results of the "data grinding.")

Experimentation

Experiments test the validity of hypotheses and involve a second major pattern of thought, termed **deduction** or **deductive logic** (also referred to as *if . . . then reasoning*). Deductive logic *proceeds from the general to the specific*. If a hypothesis is valid, then predictions based on it must be consistent with experimental results.

Events or conditions subject to change are termed **variables**. Experimentation attempts to examine the effect(s) of changing one condition or event on other conditions (**independent variable**), or events related to the phenomenon or phenomena being studied (**dependent variables**). For example, in an experiment testing the hypothesis that the tartness of green apples is a function of the maturation of the fruit, tartness would be a dependent variable and the degree of maturation would be the independent variable. An attempt must be made to control all other variables so that the relationship(s) between the independent and dependent variables is clearly identified.

An experiment must have a **control**, a basis for comparison. For instance, in addition to examining groups of apples (from the same tree) at different stages of development for degree of tartness, the experimenter should measure the tartness of the most mature apples. The mature apples would serve as a control group and provide a basis for measuring differences in tartness among the experimental groups. These differences can clearly be defined utilizing **quantitative data**, which are numerical measurements (for example, quantity of sugar per unit weight of fruit).

Note: Experiments do not *prove* the validity of a hypothesis; they are useful only in generating data concerning the consistency of hypothetical predictions versus experimental outcomes. Also, experiments must be *repeatable*, for otherwise their results could simply be relegated to chance.

Analysis and Interpretation

Objective **analysis** converts raw data into a form (statistics, graphics, etc.) that clearly describes some aspect of the phenomenon under study. **Interpretation** of the *results* of objective analysis leads to identifying meaningful relationships, or the lack of meaningful relationships, among the variables studied.

Conclusions

Conclusions are based on the analysis of data and the interpretation of results. This may involve the acceptance or rejection of the hypothesis. Any new insights into the problem are also described. Finally, in an attempt to guide further studies on the subject, additional questions and hypotheses may be discussed.

The steps involved in an investigation, as well as the results obtained, must be made available to the scientific community (generally through publication) for verification or disproof. Even a verified hypothesis is advanced to a theory or law only after much additional testing.

WHAT IS BIOLOGY?

Biology may be defined as the study of "life." It ranges in scope from the atomic and molecular structure of living matter, through its organization in individual organisms, to the interactions of organisms and groups of organisms with one another and with their environment.

The Properties of Life

No single definition delineates between living and nonliving things. Animate structures are distinguished from inanimate ones, however, by certain properties they possess.

Living things are highly organized. They exhibit a complex organization of atoms, molecules, cells, organs, and systems (Figure 1.1). A constant expenditure of energy is necessary to maintain this state of organization. Death results in disorganization.

Living systems maintain homeostasis — a chemical and physical composition different from that of their external environment. **Homeostasis** is dynamic in that living systems constantly adjust their internal environment relative to their constantly changing external environment in order to maintain stable and characteristic chemical composition. The ability of organisms to adjust to changes in their external environment is termed **acclimation.**

Living things assimilate energy from the environment, transform it, and use it to perform biological work. The total of all chemical and physical reactions involved in the acquisition, transformation, and use of energy is termed **metabolism.**

Living things can respond to environmental change (irritability). This may involve becoming *excited* by an environmental stimulus and *conducting* the effect of the stimulation from one part to another.

Living things have the capacity to reproduce themselves with an amazing degree of exactitude. In reproduction, organisms pass on their organization and function genetically.

Most living things exhibit growth and development. Growth and development are controlled by the hereditary information received from parents.

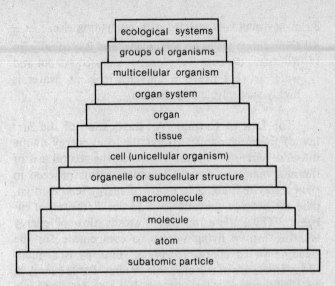

Figure 1.1. Levels of organization in biological systems.

Living things have the ability to adapt to some degree of environmental change. **Adaptation** is the evolution of heritable characteristics that render a group of organisms more fit to live and reproduce in their environment.

These characteristics are merely facets of the whole, and the whole is **synergistic** — greater than and unique from the sum of its parts.

Unifying Principles of Contemporary Biology

The principle of evolution. Evolution is considered to be the greatest unifying principle in biology. The central concept underlying this principle is that all living things have evolved and therefore share common ancestors. The existing diversity of living things is the result of their adaptation to changing environmental conditions.

The cell theory. The **cell theory** or "cell concept" states that all organisms are made up of one or more cells, which are the basic *living* units of structure and function. It also states that all cells arise from preexisting cells.

The laws of physics and chemistry. All living systems conform to the laws of physics and chemistry, which in terms of biological phenomena is summarized in four statements:

1. Energy can neither be created nor destroyed, but can only be changed (the first law of thermodynamics, or the law of conservation of energy).

2. For every action is an equal and opposite reaction. "You don't get something for nothing!"

3. Everything is connected with everything else.
4. Everything has to go someplace, and it is usually into water. Because our planet is a closed system and water is very nearly a universal solvent, water is highly susceptible to pollution.

The Law of Entropy. An implication of the first law of thermodynamics is that the total energy of the universe remains constant. However, the second law of thermodynamics adds: "All natural events proceed in such a way that concentrations of energy tend to dissipate and become unavailable to do work" (the law of entropy). Thus, there must be a steady flow of energy entering *into* a living system to compensate for the steady flow of energy *leaving* it. **Entropy** is a quantitative measure of the unusable or unavailable energy in a closed thermodynamic system. It is also described as a measure of the disorder in a closed system.

CHAPTER REVIEW

1. What is science?
2. List the hierarchy of scientific ideas in ascending order of "validity." Describe each level of the hierarchy.
3. What is the scientific method?
4. Contrast inductive logic with deductive logic. What roles do they play in scientific inquiry?
5. Experiments serve what purpose? Describe the important elements of experimentation.
6. Contrast pure science with applied science.
7. Define biology.
8. What are the characteristics of life?
9. What are the unifying principles of contemporary biology?

FUNDAMENTAL PRINCIPLES OF CHEMISTRY

Elements are the basic building blocks of all substances, including those of complex biological systems (Table 2.1). An element is a substance composed of only one kind of atom and possessing unique chemical properties. An **atom** is the most fundamental unit of matter into which a chemical element can be divided and still retain the properties characteristic of the element. Thus, atoms of different elements differ in physical and chemical properties. A **compound** is a substance composed of two or more kinds of atoms in specific ratios (H_2O, CO_2, $C_6H_{12}O_6$, for example). The most fundamental unit of a compound is a **molecule**.

Table 2.1. Atomic structure of elements most commonly found in living things.

Element	Symbol	Atomic nucleus		Number of electrons
		Number of protons	Number of neutrons[a]	
Hydrogen	H	1	0	1
Carbon	C	6	6	6
Nitrogen	N	7	14	7
Oxygen	O	8	16	8
Sodium	Na	11	23	11
Magnesium	Mg	12	24	12
Phosphorus	P	15	31	15
Sulfur	S	16	32	16
Chlorine	Cl	17	35	17
Potassium	K	19	39	19
Calcium	Ca	20	40	20
Iron	Pe	26	56	26
Iodine	I	53	127	53

[a]As per most common isotope.

MATTER AND ENERGY

All substances, including organisms and atoms, are composed of matter. **Matter** is that which exhibits the properties of *mass* and *volume*. All matter contains **energy**, which is best understood in terms of its effect upon matter. **Potential energy** is *energy available to do work*. (It is also referred to as energy of position.) **Chemical energy** is a form of potential energy contained in the bonds between atoms. A **chemical bond** is simply an energy relationship that holds the atoms within molecules together. Some chemical bonds contain more energy than others. Photosynthesis (Chapter 7), in which sunlight is used by a cell to assemble atoms and molecules into simple sugars, is an example of the storage of energy in the form of chemical bonds.

When a chemical bond is broken, energy is released. Released energy (light, heat, electricity, for example) is termed **kinetic energy** — energy of motion. Kinetic energy is *energy involved in doing work*. The heat that radiates from burning wood is an example of liberated chemical energy.

Potential and kinetic energy, in all their varied forms, are interrelated and interconvertible. The process of interconversion is called **transformation of energy**. It always occurs with some loss of usable energy — energy transformations are never 100 percent efficient. Energy unavailable to do work is a third form of energy. **Entropy** is a quantitative measure of the energy lost to a closed system.

Chemistry

Getting and expending energy is a fundamental function of life. All organisms must extract energy from

5

raw materials (sugars, starches, and other forms of food), and convert that energy into useful **work** (respiration, movement, reproduction, and so on). Consequently, living organisms are constantly involved in **energy flow** — they take in energy from outside themselves and convert it into some useful form through chemical reactions. Also through chemical reactions, a portion of that useful energy is then released to carry out life processes, after which it leaves the organism and is dispersed into the environment.

In order to understand how living things work, their **biochemistry** — the chemistry associated with living systems — must first be understood. An understanding of biochemistry is based in turn on knowledge of **organic chemistry** — the chemistry of compounds containing carbon. Finally, principles of **inorganic chemistry** — the properties and interactions of atoms and molecules — provide the basis for understanding all higher levels of chemistry. Chemistry is essential to understanding life, at least from a scientific standpoint.

THE ATOMIC THEORY

Atomic Structure

Atoms are composed of a small, dense, positively-charged central portion — the **nucleus** — surrounded by various numbers (characteristic for each element) of negatively-charged **electrons**. The nucleus contains **protons**, which are positively-charged, and usually **neutrons**, which carry no charge. Neutrons are thought to function as "packing agents," which insulate protons (like-charges) that would otherwise repel each other, and thereby lend stability to the nucleus.

The present concept of the atom involves a mathematical model (Figure 2.1) for which the probability of an electron existing at any point about the nucleus can be calculated. The location of an electron is described as an *electron cloud* because electrons randomly move about the nucleus at very high velocity. Of the atomic components, electrons are most directly involved in chemical reactions.

Atoms that have equal numbers of electrons and protons are electrically neutral (Table 2.1). Under certain conditions, however, an atom can gain or lose electrons. It consequently acquires a negative or positive charge and becomes an **ion**. When atoms interact with one another by giving up, taking on, or even sharing electrons, a chemical reaction occurs. *The exchanges and interactions of electrons among atoms form the basis of chemical reactions, and thus, of all life processes.*

The number of protons in the nucleus of an atom, which is equal to the number of electrons in a neutral atom, determines **atomic number**. Elements differ in atomic number. This is the fundamental basis for the unique chemical properties of each element. The number of protons dictates the number of electrons for an element, which in turn determines the chemical reactivity for that element.

Elements also differ in **atomic mass** — the total mass of a particular element. Atomic mass approximates the mass of all the protons and neutrons of an atom. (The contribution of electrons is negligible.)

Sometimes different atoms of the same element contain different numbers of neutrons. These atoms differ from one another in their atomic weights but not in their atomic number. Such atoms are known as **isotopes** of the element. For example,

	Hydrogen	Deuterium	Tritium
atomic mass ⟶	^{1}H	^{2}H	^{3}H
atomic number ⟶	1	1	1

The atomic mass of any element is the average of the atomic masses of the isotopes of that element weighted by their relative abundance. Most elements have several isotopic forms. Many, but not all, of the less common isotopes are radioactive. The nucleus is unstable and emits ionizing radiation as it changes to a more stable form. Isotopes have important uses in biological research and medicine as "tracers," as in radioactive isotope dating, for example. (Tracers are used to track substances as they move through biological systems.)

Energy Levels

Each electron in the electron cloud of an atom possesses a certain amount of potential energy, which is determined by its distance from the nucleus. An electron with a relatively low amount of energy is located near the nucleus, and is said to be at a low **energy level**; an electron at a higher energy level has more energy and is farther from the nucleus. Electrons at minimum energy

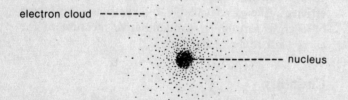

electron cloud -------

------- nucleus

Figure 2.1.   Contemporary model of atomic structure. The myriad of dots (the electron cloud) about the nucleus represents the electron distribution probability: The greater the density of dots, the greater the probability that an electron will be found in that region.

levels are said to be at **ground state**. However, if energy is absorbed by the atom, one or more electrons may jump from a lower energy level to a higher one. Such atoms are said to be in an **excited state**. When one or more electrons fall from a higher energy level to a lower one, they radiate a precise amount of energy.

An attractive force exists between negatively charged electrons and the positively charged nucleus of the atom. This force is greatest at the first energy level, and falls off successively for more distant energy levels. Consequently, ground-state electrons in the outer energy levels are more easily removed from an atom than those close to the nucleus.

Energy is temporarily captured by atoms whose ground-state electrons at outer energy levels jump to higher energy levels. It may then be released in ways that allow electrons in other atoms to make transitions also. Thus, energy can be transferred among groups of atoms by the movement of electrons from one energy level in one atom to a different energy level in another.

Electrons can jump several energy levels, depending on the amount of energy supplied, or even jump far enough from the nucleus that they escape completely from the atom. The loss of an electron (**oxidation**) gives the resulting ion a charge of $+1$ because it now has one more proton than electrons; the loss of an additional electron results in a charge of $+2$, and so on. Ions can also be formed by the gain of electrons (**reduction**), and these ions are negatively charged. The process of losing or gaining electrons is called **ionization**.

When most *biological* molecules are oxidized, electrons are removed in combination with protons rather than alone. For example, an entire hydrogen atom — one proton and one electron — might be removed. Oxidation and reduction in biological systems are frequently associated with the transfer of hydrogen protons.

MOLECULES: STRUCTURE AND FUNCTION

When atoms *react* with one another, their outer energy levels become filled, and new and larger particles called molecules are formed. **Molecules** are particles consisting of two or more atoms. The forces that hold atoms together in a molecule are known as **bonds** or **chemical bonds**. Relatively weak forces — **hydrogen bonds** — also bind molecules together.

Chemical Bonds

Chemical bonds represent energy relationships between two or more atoms. Electron orbitals are changed among the atoms involved (by filling electron vacancies in their energy levels) to achieve greater stability. These stable chemical bonds represent stored chemical energy (potential energy), with certain bonds containing more energy than others. To release this energy, bonds must be broken, and this in turn requires the investment of some energy — **dissociation energy**. When molecules dissociate, the amount of energy given off is determined by the number of bonds broken and the rate of dissociation or reassociation.

Elements differ with respect to their degree of chemical reactivity. This is a function of the outer electron configuration of their atoms. Atoms whose outer energy levels are filled (helium, neon, and argon, for example; Table 2.2) are very stable and therefore nonreactive. Most atoms, however, are reactive (have fewer than eight electrons in their outer energy levels), and tend to reach stable electron configurations by relinquishing, acquiring, or sharing electrons. The tendency of atoms to attain stable outer energy levels is, in fact, the driving force behind any chemical reaction.

Types of chemical bonds. **Ionic** or **electrovalent bonds** are the most prevalent chemical bonds found in inorganic compounds (compounds lacking carbon). In the formation of an ionic bond, one atom gives up its outermost electron(s) to one or more other atoms (Figure 2.2). When this occurs, the outermost energy level of each atom becomes stable, and simultaneously each atom becomes ionized. The positive and negative charges result in their strong mutual attraction and subsequent bond formation. However, *there is no 100*

Table 2.2. The ground state electronic configuration of elements most commonly found in living things, and three nonreactive elements.

Atomic number	Element	Number of electrons Energy levels		
		1	2	3
1	H	1		
2	He[a]	2		
6	C	2	4	
7	N	2	5	
8	O	2	6	
10	Ne[a]	2	8	
11	Na	2	8	1
12	Mg	2	8	2
15	P	2	8	5
16	S	2	8	6
17	Cl	2	8	7
18	Ar[a]	2	8	8

[a]Nonreactive elements.

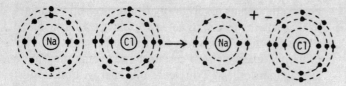

Figure 2.2.   Representation of ionic bond formation. Sodium loses the single electron in its outer electron shell to chlorine, whose outer shell is then filled. Consequently, sodium acquires a positive charge and chlorine acquires a negative charge. The difference in charges that results causes their strong mutual attraction and subsequent bond formation.

percent ionic bond. Even though one atom gives up electrons to another, this "handing over" is not complete — the donated electron may still occasionally move about the nucleus of the donor atom. In general, *those atoms with fewer than four electrons in the outer energy level tend to be oxidized; those with more than four tend to be reduced* (Table 2.2).

Bonds formed by shared pairs of electrons are known as **covalent bonds**. In a covalent bond the shared pair of electrons forms a new **molecular orbital**, that envelopes the nuclei of both atoms. In such a bond, each electron spends part of its time around one nucleus and part around the other. Thus, an electron sharing both nuclei completes the outer energy level and neutralizes the nuclear charge (Figure 2.3). Atoms that need to *gain* electrons to achieve a filled, and therefore stable, outer energy level have a strong tendency to form covalent bonds.

The capacity of carbon atoms to form covalent bonds is extremely important in living systems. A carbon atom has four electrons in its outer energy level (Figure 2.3). It can share each of those electrons with another atom, forming covalent bonds with as many as four other different atoms (most frequently hydrogen, oxygen, and nitrogen) or with other carbon atoms. The tendency of carbon atoms to form covalent bonds with other carbon atoms gives rise to the large molecules that form the structures of living organisms, and that participate in essential life processes. Consequently, carbon is referred to as "the building block of life." (Silicon has

similar properties, and it is possible that life on other planets are based on silicon.)

The electrons in covalent bonds are not always shared equally between the atoms involved. For example, if the nucleus of one kind of atom has a greater attractive force for electrons than does the nucleus of another kind, the shared electrons tend to spend more time around the nucleus with the greater attraction. The atom with which the electrons spend more time has a slightly negative charge (whereas the other atom has a slightly positive charge) because its nuclear charge is not entirely neutralized. Covalent bonds in which electrons are shared unequally are known as **polar covalent bonds**, and molecules containing these bonds are called **polar molecules**. Polar molecules often contain oxygen atoms to which electrons are strongly attracted. The polar properties of many oxygen-containing molecules have important consequences for living systems. For example, many of the special properties of water (H_2O), upon which life depends, are derived largely from its polar nature.

Ionic bonds are weaker than covalent bonds. The dissociation of an ionically-bonded molecule merely involves overcoming the electrostatic attraction between positive and negative ions. When such molecules dissociate, no further electron exchange is required. With covalent bonds, however, the outermost energy level of each atom is satisfied only as long as the shared electrons move about both nuclei. A stronger chemical bond results because the atoms must remain in close proximity. It is difficult to separate one covalently bonded atom from another if they are forced to assume unstable outer electron configurations.

Chemical Reactions

Chemical reactions involve the making and breaking of chemical bonds, and subsequent energy transformations.

The collision theory. The **collision theory** of chemical reactions is derived from the idea that all atoms, molecules, and ions in any system are in constant motion (**Brownian motion**). Particles that are to interact chemically must first come into contact for electron exchanges or rearrangements to be possible. The collision of any two particles is considered to be a completely random event.

Not every collision between particles produces a chemical interaction. First, *the average velocity of the particles determines what percentage of collisions are successful for any given kind of reactants*. The more rapidly the particles travel, the more likely they will

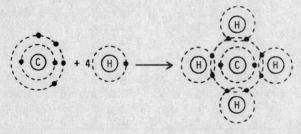

Figure 2.3.   Covalent bonding.

yield successful collisions. Second, *particles of each element or compound have their own minimum, or threshold, energy requirements for successful interaction*. This energy is usually referred to in terms of particle velocity (a manifestation of kinetic energy). Below a certain velocity, collisions for a given element will probably not be successful. If the average kinetic energy of a system is increased, the number of successful collisions will also generally be increased. Third, *molecular geometry of particles plays a role in determining whether or not a collision is successful*. If a molecule collides with an atom or another molecule in such a way that the reactive portion of the molecule is not exposed to the other particle, *no reaction will occur*, even if the particles possess the proper amount of kinetic energy.

Activation energy. Activation energy (E_a) is the initial input of energy that must be supplied from an outside source before chemical reactions can begin (Figure 2.4). In other words, it is the minimum kinetic energy required by any system of particles for any successful chemical reaction. If the average energy is below this minimum, the reaction will proceed slowly or not at all. If the average is above the minimum, the reaction will proceed more rapidly. **A catalyst** is any substance that lowers the activation energy, thereby allowing a given chemical reaction to proceed more rapidly. Although it may be temporarily changed during the reaction, a catalyst is unchanged when the reaction is done. **Enzymes** are biological catalysts.[1] Living systems use enzymes to "hold" large molecules in specific positions, which induces internal rearrangements, cleavage, or union with other molecules.

Chemical reactions can be discussed only in terms of probability. The rate of a chemical reaction is influenced by several factors (concentration of reactants, catalysts, temperature, pH, among others) that increase or decrease the probability that collisions between particles will be successful.

The energy exchanged in chemical reactions is a critical factor in living systems. In Figures 2.4 and 2.5, the energy exchanged (at constant pressure) is symbolized by ΔE and is called **enthalpy**. Enthalpy is the difference in potential energy between the products and the reactants, an important property of any system. If ΔE is negative, the reaction is **exergonic** and energy is given off (Figure 2.4). The energy liberated is referred to as **free energy**. If ΔE is positive, the reaction is **endergonic** and energy is absorbed by the system (Figure 2.5).

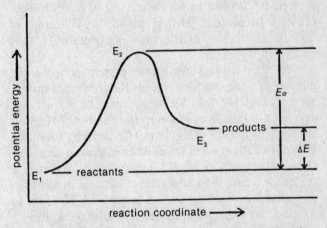

Figure 2.5. A potential energy diagram for an endergonic reaction.

Reversible and irreversible chemical reactions. Chemical reactions are **reversible**: a reaction can go in either direction. For example,

$$2H^+ + O^{-2} \longrightarrow H_2O$$

$$H_2O \longrightarrow 2H^+ + O^{-2}$$

Thus,

$$2H^+ + O^{-2} \rightleftharpoons H_2O$$

Note that the arrows are of equal length, which indicates that the forward reaction occurs just as readily as the reverse reaction.

In principle, *all* chemical reactions are reversible. There is no chemical reaction known that, under suitable conditions, cannot proceed (however slowly) in the reverse direction. Yet under ordinary conditions some reactions are far less reversible than others. Reactions whose reversibility is barely detectable are said to be **irreversible**.

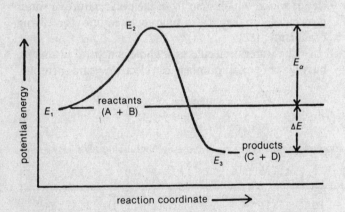

Figure 2.4. The potential energy change during a chemical (exergonic) reaction. E_1 is the potential energy of the reactants. E_2 is the energy level of the reactants necessary to initiate the reaction. Thus, the difference between E_1 and E_2 is the activation energy (E_a). E_3 is the potential energy of the products. The difference between E_2 and E_3 is the enthalpy (ΔE) of the system — the amount of energy stored or released from the chemical reaction.

Chemical equilibrium. Within a certain period of time, reactions reach a state of equilibrium (Figure 2.6). When this condition is reached, the proportion of reactants in relation to products remains the same. Notice that this does *not* mean the amounts of reactant and of product are necessarily equal. *Chemical equilibrium is established when the forward rate of any chemical reaction is equal to the reverse rate.*

Acids, bases, and the pH scale. Energy transfer in chemical reactions may be discussed in terms of the liberation and uptake of protons (hydrogen ions – H⁺). For example, in the dissociation of water (H_2O)

$$HOH \rightleftharpoons H^+ + OH^-$$

When molecules dissociate, a certain amount of energy is given off, depending on: the number of bonds broken (*number* of protons [H⁺] given up), and the rate of dissociation or reassociation (*how fast* protons [H⁺] are given up).

A solution acquires **acidic** properties when the number of hydrogen ions exceeds the number of proton acceptors (OH⁻ or hydroxyl ions of water, for example). (*Note:* In pure water, the number of H⁺ ions equals the number of OH⁻ ions.) Conversely, a solution is **basic (alkaline)** when the number of proton acceptors exceeds the number of H⁺ ions. Thus, an acid is a substance whose molecules release H⁺ ions in solution, and a base is a substance that bonds with H⁺ ions in solution. Strong acids and bases are substances, like

$$HCl \rightleftharpoons H^+ + Cl^- \text{ (hydrochloric acid)}$$

$$NaOH \rightleftharpoons Na^+ + OH^- \text{ (sodium hydroxide)}$$

that ionize (dissociate) completely in water. Weak acids and bases are those that ionize only slightly in water.

Degrees of acidity are defined in terms of the **pH scale** (Figure 2.7). At pH 7.0, the concentration of H⁺ and OH⁻ ions is exactly the same (as they are in pure water). This is a neutral state. *Any pH below 7.0 is acidic, and any pH above 7.0 is basic.* A difference of one pH represents a tenfold difference in the concentration of H⁺ ions. Most chemical reactions of living systems take place within a narrow range of pH that hovers around neutrality. The pH of a solution can be stabilized with **buffers** — substances that can absorb or release hydrogen ions.

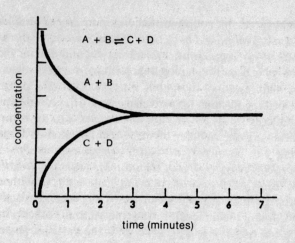

Figure 2.6. Changes in concentration over a period of time for a completely reversible reaction. Note that an equilibrium is established about 3½ min after the reaction has begun. At this point, the concentration of reactants equals the concentration of products.

The Chemistry of Water

Water, the most common liquid on the earth's surface, is the major component, by weight, of all living things. The living cell is approximately 90 percent water. Life as we know it occurs in a water medium and could not exist otherwise. The chemical and physical properties of water, which stem from the polar nature of water molecules, make it a unique medium for living organisms.

The water molecule as a whole is neutral in charge, having an equal number of electrons and protons.

FIgure 2.7. The pH scale. Values range from 0 through 14. The actual concentration of hydrogen ions is expressed in moles per liter. (*Note:* In this case, one mole represents 6.02 × 10²³ ions of hydrogen.) Some pH values of solutions in the human body are indicated.

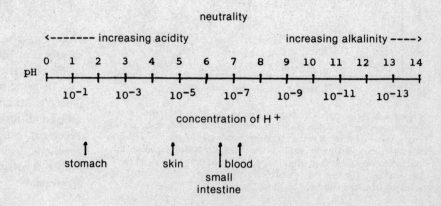

Because of the strong attraction of the oxygen nucleus for electrons, however, the shared electrons of the covalent bonds spend more time around the oxygen nucleus. Thus, the region near each hydrogen nucleus is a weakly positive zone while the region near the oxygen nucleus is weakly negative (Figure 2.8). This positive region gives the water molecule the capacity to form a special kind of bond — the **hydrogen bond**. A hydrogen bond is formed when one charged region comes close to an oppositely charged region of another molecule. Every water molecule can establish hydrogen bonds with four other molecules. As a result of hydrogen bonding, water is highly cohesive, a good solvent, has high specific heat, and is less dense as a solid than as a liquid.

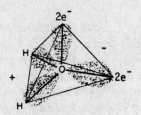

Figure 2.8. An orbital model of the water molecule showing its polar nature.

CHAPTER REVIEW

1. Define and contrast the following terms:
   a. element and atom
   b. compound and molecule
   c. atomic number and atomic mass
   d. atom, ion and isotope
   e. oxidation and reduction
2. What is matter?
3. Describe the three principle forms of energy.
4. Describe the contemporary model of atomic structure.
5. Explain how atoms absorb and radiate energy. Describe how energy can be transferred among groups of atoms.
6. Describe the various forces that hold atoms and molecules together.
7. Define and discuss the following terms within the context of chemical reactions: activation energy, catalyst, and chemical equilibrium.
8. Define the following terms: pH, pH scale, acidic solution, and basic solution.
9. Discuss what it is about water that results in its unique chemical and physical characteristics.

----

[1]Enzymes are discussed more fully in Chapter 3, under "Enzyme Physiology."

THE CHEMISTRY OF LIFE

Because organic chemistry deals only with the structure and function of carbon-containing compounds, it might seem somewhat more restricted than inorganic chemistry. However, organic chemistry is a much broader field because *carbon* is capable of forming more varieties of compounds with more different elements than almost any other atom in the periodic table. This fact is well demonstrated in living things.

Water makes up from 50 to 95 percent of an organism, and ions (sodium, Na^+, and potassium, K^+, for example), account for no more than 1 percent. The remainder of a living system, chemically speaking, is composed almost entirely of organic molecules (Figure 3.1).

Organic chemistry (including biochemistry) involves the principles of chemical reactions, bond formation, and energy exchange. There is nothing fundamentally different about the chemical reactions within living organisms and those in nonliving systems. Reactions within living systems, however, are generally more complex and involve more individual steps and specialized energy sources than those occurring in inorganic or nonliving systems. *This greater complexity is*

important, for it imparts to living systems their unique abilities to organize and regulate chemical reactions in a way that produces the characteristics of life.

THE CARBON BACKBONE

Two important features of carbon result in its being the backbone of all organic compounds:

1. A carbon atom can form four covalent bonds with as many as four different atoms (Figure 3.2).

2. More important biologically, carbon atoms can form bonds with each other (Figure 3.3).

In general, *an organic molecule derives its overall shape from the arrangement of carbon atoms that form the skeleton of the molecule* (Figures 3.3 and 3.4). Its shape in turn determines many of its properties and functions within living systems. For example, **isomers** are molecules that have the same chemical formula, but differ in configuration and subsequently in their chemical characteristics.

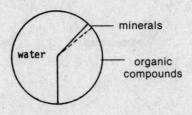

Figure 3.1. The composition of cells.

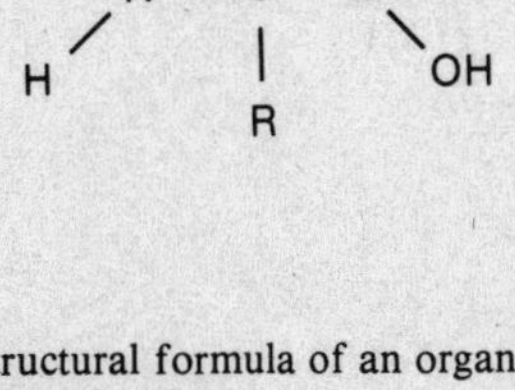

Figure 3.2. The structural formula of an organic molecule showing the bonding versatility of carbon.

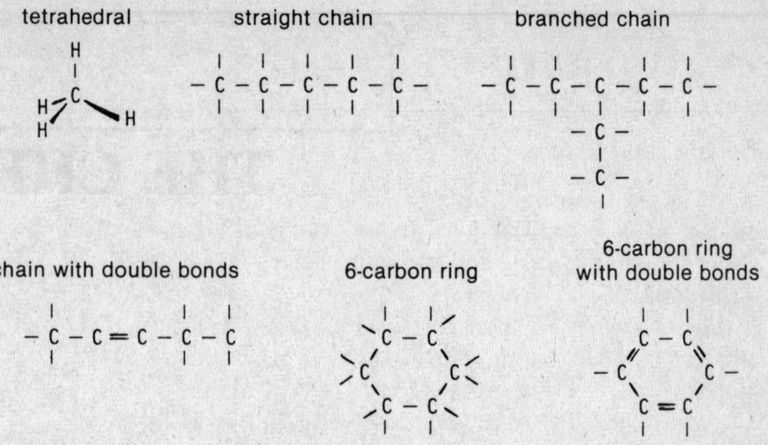

tetrahedral   straight chain   branched chain

Figure 3.3. Various forms of structural conformations that result from carbon-carbon bonding.

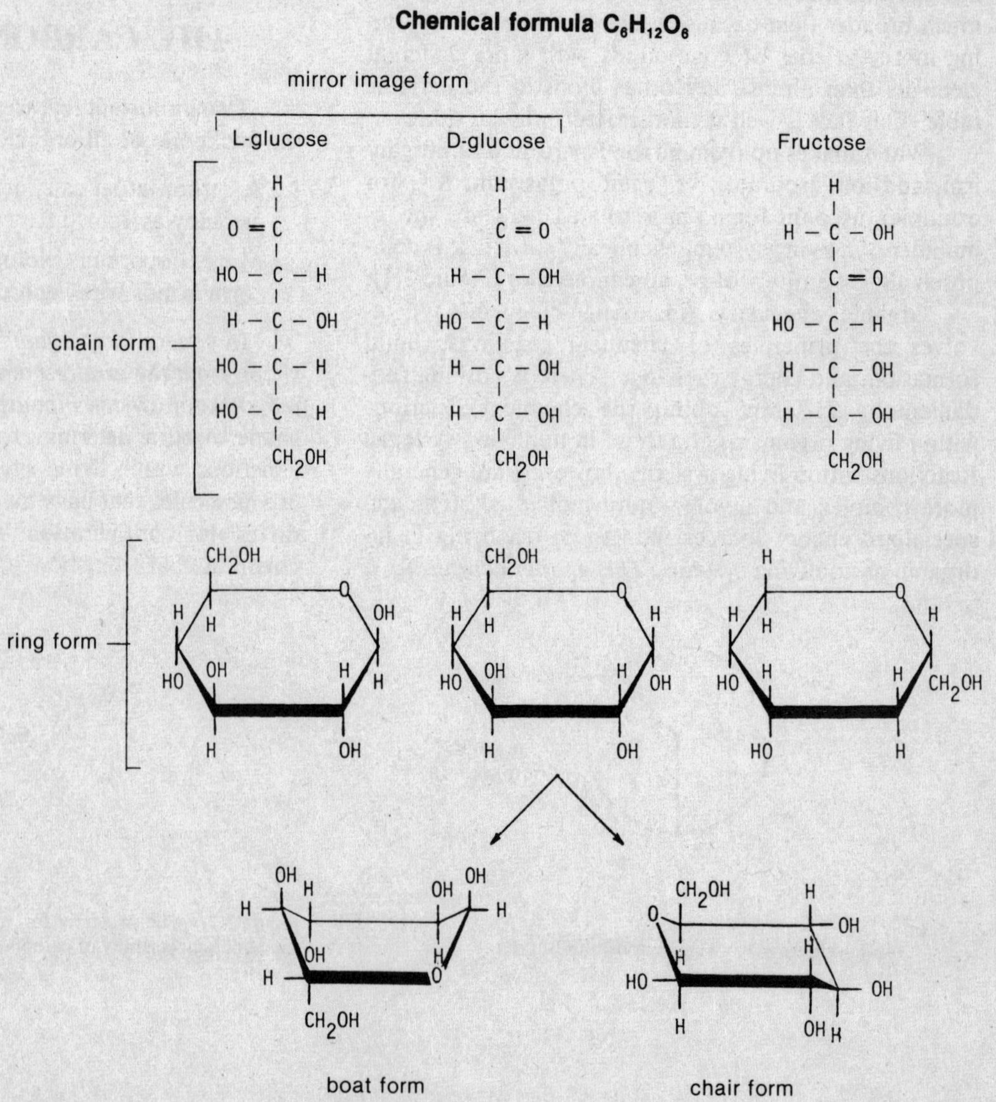

chain with double bonds   6-carbon ring   6-carbon ring with double bonds

Chemical formula C₆H₁₂O₆

mirror image form

L-glucose   D-glucose   Fructose

chain form

Figure 3.4. Various forms of structural isomerism of organic molecules. Each of the six-carbon sugars has the same molecular formula, C₆H₁₂O₆; thus each is an isomer of the others.

ring form

boat form   chair form

CLASSES OF MOLECULES IN LIVING SYSTEMS

Carbohydrates

Carbohydrates are organic compounds composed of carbon, hydrogen and oxygen. Carbohydrates generally contain hydrogen and oxygen in a 2:1 ratio. Thus, a generalized carbohydrate formula can be written as $[C(H_2O)]_n$, where n refers to the numbers of this unit composing the molecule. For example, the formula for glucose is $C_6H_{12}O_6$ where $n = 6$.

Carbohydrates serve two essential functions in living organisms:

1. Carbohydrates are the principal *energy-storage molecules* in most organisms.

   starch – energy storage molecule for plants

   glycogen – energy storage molecule for animals

2. Carbohydrates are essential *structural components* in many organisms, particularly plants.

   cellulose – primary component of the cell wall in plants

   chitin – major component of the exoskeleton of invertebrate animals (e.g., insects and crustaceans)

Monosaccharides. The fundamental building blocks of carbohydrates are **monosaccharides** – simple five- or six-carbon sugars. These sugars, particularly glucose, are the principal sources of energy for most organisms. In the process of *photosynthesis*, cyanobacteria and the green cells of plants capture energy from sunlight (through *chlorophyll*), and use it to combine carbon dioxide and water into sugar molecules (with oxygen as a by-product):

$$CO_2 + H_2O \xrightarrow[\text{sunlight}]{\text{chlorophyll}} (CH_2O)_n + O_2$$

Photosynthesis is an example of an **anabolic chemical pathway** in which captured energy is stored in the chemical bonds of the sugar molecules.[1]

In another series of reactions, living cells break down (oxidize) the sugar molecules, releasing carbon dioxide, water and energy:

$$(CH_2O)_n + O_2 \longrightarrow CO_2 + H_2O + energy$$

This is an example of a **catabolic chemical pathway** in which larger molecules are broken down into smaller ones, and energy is released as the result of the breaking of chemical bonds. Other organic substances also release their stored energy when they are oxidized (for example, when wood, gasoline, or alcohol are burned for fuel).

Disaccharides. Two simple sugars joined together (e.g., glucose and glucose, or glucose and fructose) are **disaccharides**. Disaccharides are formed by the removal of a molecule of water from two monosaccharide molecules — **condensation reaction (dehydration synthesis)**. The result is the formation of a covalent bond linking two monosaccharide molecules (Figure 3.5). Condensation reactions require energy. The opposite chemical reaction — **hydrolysis** — releases energy. Hydrolysis is the addition of a molecule of water to a disaccharide, causing it to split into two monosaccharides. (Most reactions in the digestive tract are hydrolytic reactions.)

Polysaccharides. Polysaccharides are large carbohydrate molecules composed of monosaccharides linked together in long chains. These large, complex molecules are called **macromolecules**. For many organisms polysaccharides serve as energy storage (for example, in the chemical bonds of starch and glycogen). They must be hydrolyzed to monosaccharides or dissacharides before they can be used as energy sources or be transported through living systems. Polysaccharides (e.g., cellulose and chitin) are also involved in the structure of organisms.

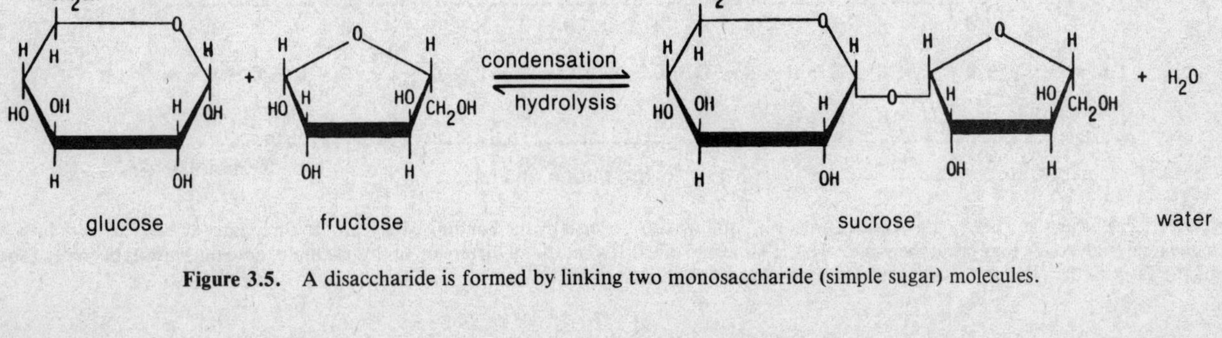

Figure 3.5. A disaccharide is formed by linking two monosaccharide (simple sugar) molecules.

Lipids

Lipids are organic chemical compounds that include the fats, oils, sterols, and waxes. They have two principal characteristics:

1. They are nonpolar and so are generally insoluble in water.
2. They contain a larger proportion of carbon-hydrogen bonds than any other organic compounds except hydrocarbons.

Because lipids have a large number of carbon-hydrogen bonds, they store more energy than most other organic compounds. Lipids differ from carbohydrates in that the ratio of hydrogen to oxygen is far greater than 2:1.

Fats and Oils. **Fats** and **oils** are the chief forms in which lipids are stored in cells, having been synthesized by cells from sugars. Their most prevalent form is that of **triglycerides** — molecules composed of a **glycerol** and three **fatty acids** (Figure 3.6). The glycerol (an alcohol) and fatty acids (unbranched carbon skeletons to which numerous hydrogens are attached) are joined by enzyme-mediated condensation reactions. The structural characteristics of a fat or oil are determined by the length of the carbon chains in the fatty acids and by whether the acids are **saturated** or **unsaturated**. In saturated fatty acids every carbon atom in the chain (except the last one) holds two hydrogen atoms, which completes (saturates) the bonding possibilities of the carbon atom. Unsaturated fatty acids contain carbon atoms joined by double bonds, and consequently do not contain the maximum number of hydrogens.

Fats are an organism's most concentrated source of biologically usable energy (potential energy). Most fats provide up to twice as many calories per gram as do carbohydrates. The chemical reasons for this are evident from a comparison of molecular formulas. Fats contain considerably more hydrogens per molecule than carbohydrates. These hydrogen atoms supply electrons for the energy-releasing chemical processes in cells, particularly oxidation, the basic process in energy release. The oxidation of fats can create considerably more total energy than oxidation of even the large carbohydrates because fats have hydrogens where carbohydrates have some oxygens. Relative to fats, therefore, carbohydrates are partially oxidized and thus yield less energy upon completion of the oxidation process.

Phospholipids. Phospholipids are lipids found in biological membranes, and are composed of two fatty acids joined to a **glycerol phosphate** (Figure 3.7). As a result of the glycerol phospate end having a net negative charge, they are polar molecules. (Phospholipids are discussed more fully in Chapter 4.)

Other lipids. Other biologically important lipids include

glycolipids — function as one kind of site for chemical recognition between cells and substances in their external environment

steroids — include *cholesterol, sex hormones* (e.g., estrogen and testosterone), and *cortisol* (secreted by the adrenal cortex)

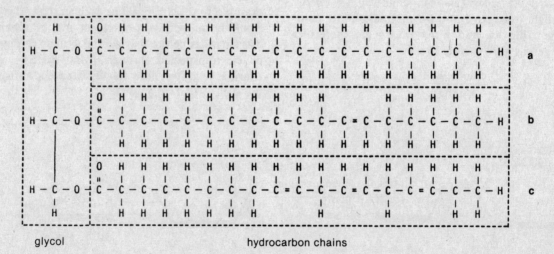

glycol   hydrocarbon chains

Figure 3.6. The basic structure of triglycerides (fats, oils, and waxes), formed by the bonding of glycerol to three fatty acids. The three fatty acids that make up a triglyceride may be all the same, two the same and one different, or all different. In this example, note the various degrees of saturation of the fatty acids: (a) saturated, (b) unsaturated, and (c) polyunsaturated.

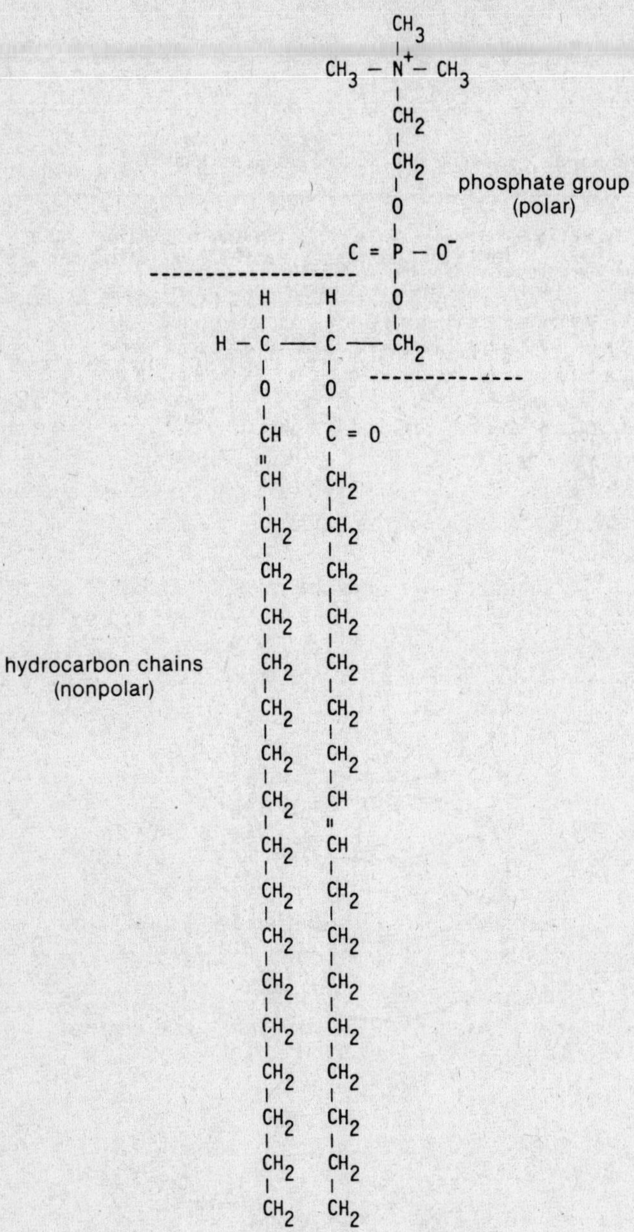

```
              CH3
               |
     CH3 — N+ — CH3
               |
              CH2
               |
              CH2       phosphate group
               |             (polar)
               0
               |
          C = P — 0-
               |
      H    H    0
      |    |    |
  H — C —— C —— CH2
      |    |
      0    0
      |    |
      CH   C = 0
      "    |
      CH   CH2
      |    |
      CH2  CH2
      |    |
      CH2  CH2
      |    |
      CH2  CH2
      |    |
      CH2  CH2
      |    |
      CH2  CH2
      |    |
      CH2  CH
      |    "
      CH2  CH
      |    |
      CH2  CH2
      |    |
      CH2  CH2
      |    |
      CH2  CH2
      |    |
      CH2  CH2
      |    |
      CH2  CH2
      |    |
      CH2  CH2
      |    |
      CH2  CH2
      |    |
      CH2  CH2
```
hydrocarbon chains (nonpolar)

Figure 3.7. The structure of a phospholipid.

terpenes — contribute to the structures of *chlorophyll* and *carotenoid* pigment molecules, which function in trapping light energy (Chapter 7)

waxes — found as protective coatings on skin, feathers and fur, on the exoskeletons of many insects, and on the leaves of fruit and plants

Proteins

Proteins are complex organic compounds composed of one or more polypeptide chains, each made up of many (about 100 or more) amino acids linked together by peptide bonds. The fundamental building blocks of protein are **amino acids** (Figure 3.8). They are nitrogen-containing compounds with an **amino group** ($-NH_2$), a **carboxyl group** ($-COOH$), a hydrogen atom, and some distinct atom or group of atoms designated as the **R group**. Amino groups give basic properties to amino acids, while carboxyl groups give acidic properties. The number and arrangement of atoms that compose the R group makes one amino acid different from another. These differences determine the chemical properties of the individual amino acids, and as a consequence, the properties of proteins as well. From twenty different amino acids, all the proteins known to exist in plants and animals are constructed.

Amino acids in a protein are linked end-to-end to form long chains. When the amino end of one amino acid is linked to the carboxyl end of another by the removal of water (condensation reaction or dehydration synthesis), a **peptide bond** is formed between the two amino acids (Figure 3.9). A molecule formed by the linking of many amino acids is called a **polypeptide**. *The variety found among proteins is the result of the types of amino acids constituting each, and the sequence in which these types are arranged.*

Structural types of proteins. The sequence of amino acids formed from a series of covalent peptide bonds may be considered the most fundamental level of organization of proteins — the **primary structure** (Figure 3.10a). This sequence of amino acids determines the structural characteristics of each type of protein molecule and thus its biological function.

The **secondary structure** is a polypeptide chain folded into a simple pattern: an **alpha helix** or a **beta sheet** (Figure 3.10b). The secondary structure is held in shape by hydrogen bonds. Although quite weak individually, many hydrogen bonds reinforce each other to produce a relatively stable structure.

Protein that has the alpha helix structure (hair and muscle protein, for example) is *elastic*. The protein is stretchable because the hydrogen bonds that hold adjacent coils together can break and reform easily.

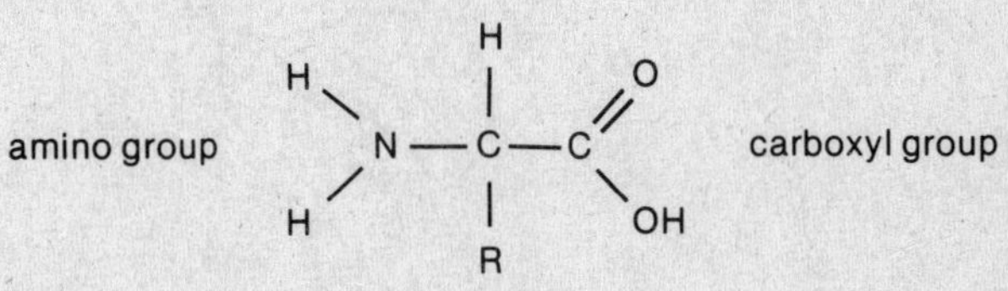

```
                      H
              H       |      O
               \      |     //
  amino group   N — C — C
               /      |     \
              H       |      OH
                      R
```
amino group   carboxyl group

Figure 3.8. The generalized structure of amino acids which includes an amino group ($-NH_2$), a carboxyl group ($-COOH$), a hydrogen atom, and some distinct functional (R) group.

Figure 3.9. Peptide bond formation. In most cellular environments, the amino and carboxyl groups are ionized.

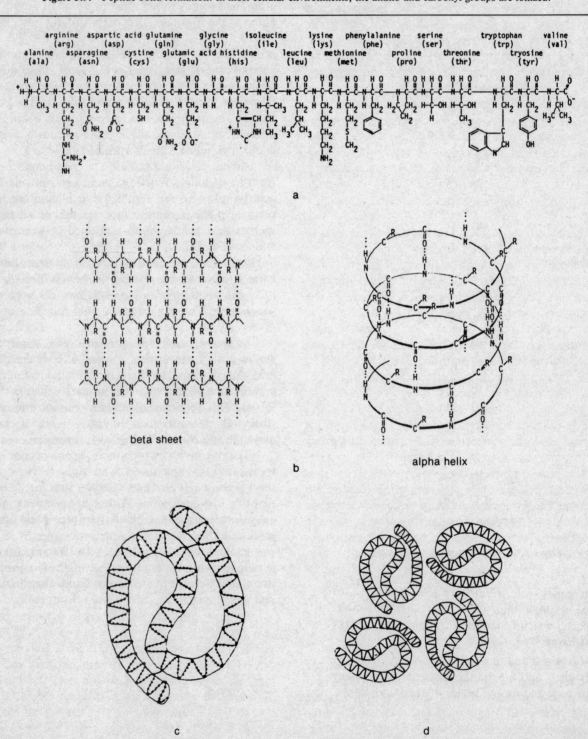

a

beta sheet

b

alpha helix

c   d

Figure 3.10. Levels of protein organization: (a) primary structure (a linear sequence of the 20 naturally occurring amino acids), (b) secondary structures, (c) tertiary structure and (d) quaternary structure (the hydrogen bonds connecting the tertiary structures are not shown). (Dotted lines represent the hydrogen bonds that form between polypeptide chains.)

The beta sheet structure consists of extended polypeptide chains lined up in parallel and linked by hydrogen bonds or covalent bonds. Protein of a beta sheet structure is smooth and flexible, but *not* elastic (e.g., silk).

Tertiary structures (Figure 3.10c) result from a secondary structure folding back upon itself. These structures form spontaneously from the attractions and repulsions among amino acids with different charges on their R groups. Proteins with tertiary structures are called **globular proteins.** Important examples of globular proteins include enzymes and antibodies.

The fourth level of protein architecture, the **quaternary structure** (Figure 3.10d), results from two or more polypeptide chains interacting to form a functional protein. For example, hemoglobin, the oxygen-carrying pigment of red blood cells, comprises four polypeptide chains held together by hydrogen bonds.

Protein function. Of all macromolecules, proteins are involved in the greatest variety of roles. Many proteins serve as enzymes, which are essential for catalyzing all biochemical reactions. In addition, proteins are important structural components in cell membranes, skeletal structures, hair, nails, muscle, and connective tissue (such as tendons and ligaments). Under extreme conditions, energy can be obtained from the breakdown (oxidation) of proteins. When amino acids are broken down, ammonia (a toxic waste product) is formed, which is removed to form urea to be excreted as urine by mammals.

The chemical properties of each kind of protein molecule are a consequence of its structure. Under less than optimal conditions – in extremes of pH or temperature, for example – the weak forces (hydrogen bonds) contributing to the structure of the protein are overcome. It subsequently unwinds or unfolds. Because of the resulting change in structure, the protein is said to be **denatured**. Up to a limit, such a process is reversible. At certain extremes, denaturation becomes effectively irreversible because chemical bonds do not readily reform. The denatured state of white or albumin of a hard-boiled egg is irreversible, for example.

Nucleotides and Nucleic Acids

As illustrated in Figure 3.11a, **nucleotides** are molecules composed of a phospate, a five carbon sugar (either ribose or deoxyribose), and a purine (double ring) or pyrimidine (single ring) base. Nucleotides serve three important functions in living systems. These functions are discussed more fully in later chapters, particularly in Chapters 5, 6, and 7:

1. They are involved in the direct release of energy to do biological work. The nucleotide ATP is called the energy currency of the cell (Figure 3.11b).
2. As energy carriers they function as coenzymes (NAD$^+$ for example, in Figure 3.11c).
3. As shown in Figure 3.11d, they form the basic building blocks of nucleic acids (DNA and RNA) and, subsequently, of genetic systems.

Nucleic acids are long single- or double-stranded chains of nucleotides.

ENZYME PHYSIOLOGY

Enzymes deserve special attention because of their involvement in virtually *all* biochemical reactions. Without these specialized proteins life as we know it could not exist. The reason for this lies in the nature of their function.

Enzymes determine *the kinds of biochemical reactions that will occur, and the rate at which those reactions will occur.* They catatlyze biochemical reactions by forming temporary associations with the molecule(s) — the **substrate(s)** — that will undergo chemical change. This temporary association weakens chemical bonds and permits new ones to form more easily. Thus, in the presence of an enzyme, very little energy is needed to initiate a reaction, and the reaction proceeds more rapidly. Minsicule amounts of enzyme are effective: A single enzyme molecule can catalyze tens of thousands of identical molecules in a second. Some can even catalyze up to 10,000,000 molecules of substrate per second for short periods of time!

Enzymes are **substrate specific;** they are limited to catalyzing specific chemical reactions involving specific substrates. This substrate specificity and the catalytic properties of an enzyme are a consequence of its three dimensional (tertiary) structure. On the surface of an enzyme molecule there is a region, the **active (catalytic) site**, into which one or more specific substrate molecules fit (Figure 3.12).

Enzymes characteristically work in *series*, in *conjunction* with other enzymes or coenzymes. The hydrolysis of starch to glucose, for example, requires coenzymes. **Coenzymes** are nonprotein organic molecules (e.g., vitamins) that often function as electron carriers. Several kinds of coenzymes exist in any given cell, each able to hold one or more electrons at slightly different energy levels. A coenzyme can accept electrons at one particular step in a reaction series, then release them at another step in the same series or in a different series.

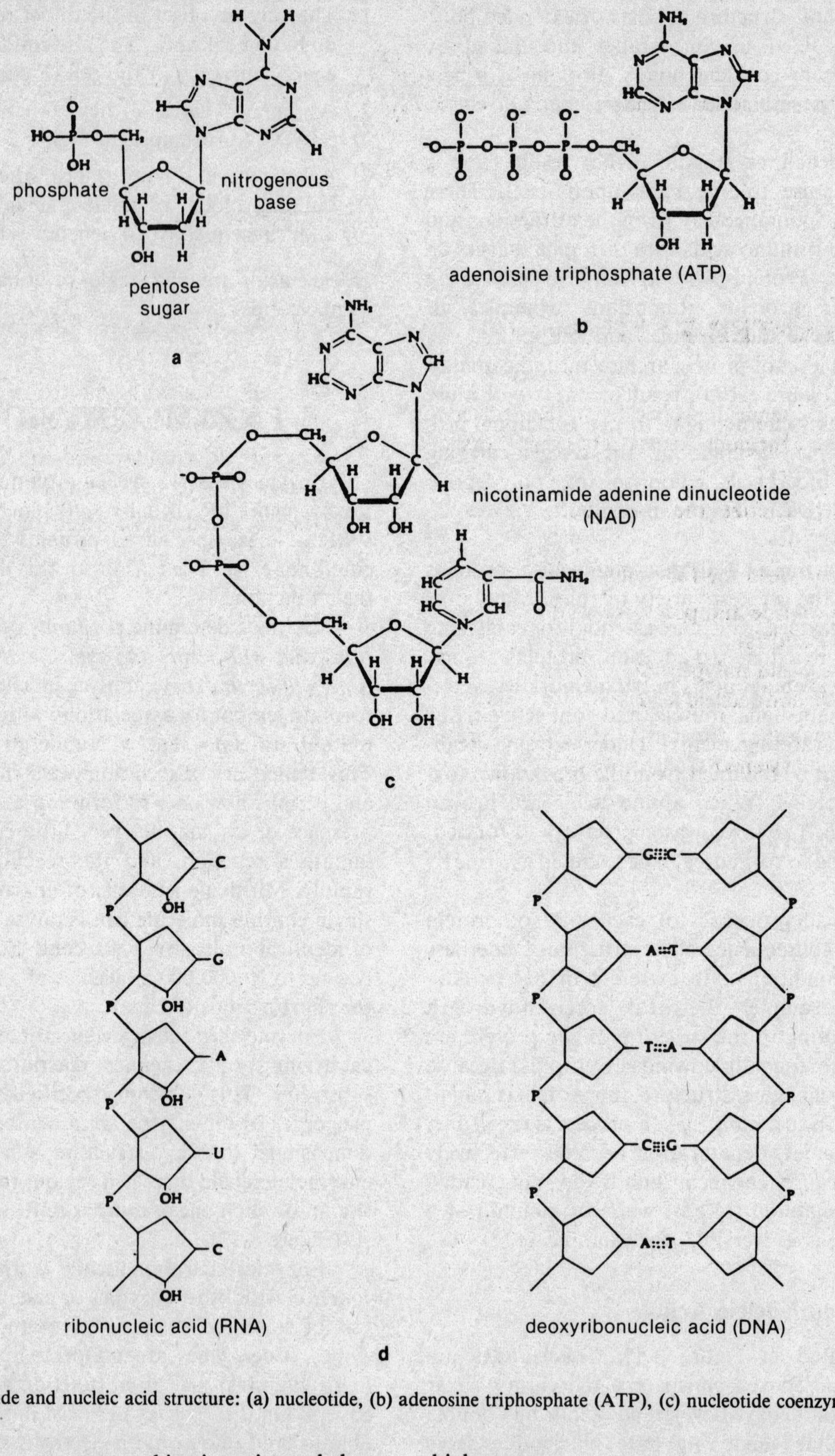

Figure 3.11. Nucleotide and nucleic acid structure: (a) nucleotide, (b) adenosine triphosphate (ATP), (c) nucleotide coenzyme, and (d) nucleic acids.

This combination — enzymes working in series and electron carrier molecules holding electrons at different energy levels — enables the cell to capture energy efficiently as electrons move from lower energy levels to higher ones.

Factors influencing enzyme-mediated chemical reactions include enzyme and substrate concentration, temperature, pH, and enzyme inhibitors.

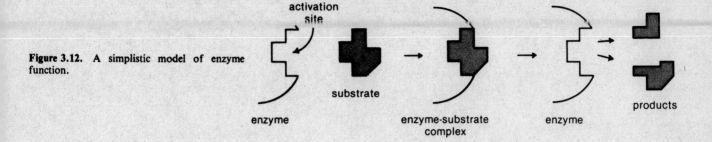

Figure 3.12. A simplistic model of enzyme function.

CHAPTER REVIEW

1. How do the chemical properties of carbon contribute to the enormous variety and complexity of organic molecules?

2. List and characterize the four main classes of organic molecules.

3. Distinguish between the following terms:
   a. monosaccharide and polysaccharide
   b. fatty acid and fat
   c. amino acid and polypeptide
   d. nucleotide and nucleic acid

4. What two essential functions do carbohydrates serve in living systems? Give examples.

5. Contrast condensation (dehydration) reactions with hydrolytic reactions.

6. List and describe the biological importance of the major kinds of lipid molecules.

7. Why are fats said to be the most concentrated sources of potential energy in living systems?

8. Describe the different levels of protein structure.

9. What are three important functions of nucleotides in living systems?

10. As biological catalysts, how do enzymes control biochemical reactions?

11. Define the following terms: substrate specific, active site and coenzymes.

¹Photosynthesis is discussed in detail in Chapter 7.

CELL BIOLOGY

THE "CELL CONCEPT"

The "cell concept" states that the cell is the fundamental unit of life. It consists of the following four propositions:

1. Cells are the *structural units* of all organisms. All organisms are composed of one or more cells.

2. Cells are the *functional (and dysfunctional) units* of all organisms. Whatever is right or wrong with an organism ultimately depends on the individual cells that make up the organism.

3. Cells *arise only from preexisting cells*. Thus, organisms arise only from preexisting organisms (**biogenesis**), and their reproduction is in some way a consequence of the reproduction of their cells.

4. Cells contain *hereditary material* through which specific characteristics are passed on from parent cell to daughter cell. The form and function of any organism (unicellular or multicellular) rests ultimately in the hereditary information (DNA) carried by its cell(s).

THE ORIGIN OF LIFE AND THE EVOLUTION OF CELLS

Chemical Evolution

The age of the Earth is estimated to be about 4.6 billion years. Prior to the appearance of life, this planet underwent over a billion years of chemical and geological development (Figure 4.1). This atmosphere did not possess oxygen, but did possess the constituents of organic molecules (ammonia [NH_3], methane [CH_4],

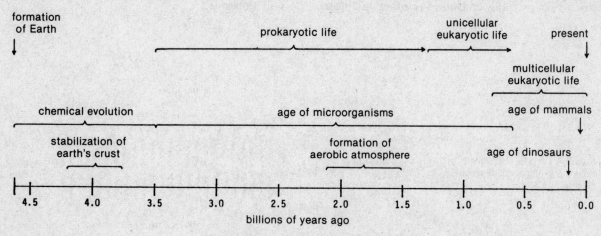

Figure 4.1. The origin of life.

hydrogen [H_2], and water [H_2O], among others). In addition, primitive Earth abounded with energy (radiation, heat, and electricity). As a consequence the atmospheric components reacted and formed simple organic molecules. Precipitation leached these molecules and some of the inorganic compounds from the atmosphere into shallow seas, lakes, and ponds, where they collected and concentrated to form a "chemical soup." The abundant energy present drove chemical reactions in the chemical soup, and resulted in the synthesis of more complex organic molecules.

$$\text{ancient atmosphere} + \text{energy} \longrightarrow \text{complex organic molecules}$$
$$(NH_4, CH_4, H_2, H_2O) \quad \text{(heat, radiation, electrical energy)} \quad \text{(e.g., phospholipids)}$$

Origin of Cells

It is theorized that cells arose from spheres bounded by a double layer of phospholipids. Referring to Figure 4.2, note that the polar ends of the phospholipid molecules are oriented toward water molecules because of mutual electrostatic attraction. The nonpolar ends simultaneously are oriented toward each other. Lipid bilayers probably served as a membrane to separate the internal chemical environment of lipid spheres from their external environment. The relatively stable internal environment subsequently permitted complex chemical reactions to occur that could not have in the relatively variable external one.

Life is estimated to have begun about 3.5 billion years ago. The oldest known fossils, resembling bacteria, are thought to be about 3.5 billion years old. These first cells probably consumed or otherwise incorporated organic molecules abundant in their environment, much in the way our cells (animal cells) use sugars, fatty acids, and proteins to meet energy needs. Organisms dependent upon external sources for organic molecules, for both nutrients and structural components, are classified as **heterotrophs** (animals, fungi, and many unicellular organisms). It is believed that as primitive cells proliferated over millions of years, they began to deplete their environment of organic molecules, and eventually were forced to compete for the dwindling supply. Cells that most efficiently exploited the limited resources available were the most likely to survive.

It is speculated that about two billion years ago, unicellular **autotrophs** evolved that were capable of manufacturing their own energy-rich organic molecules. Autotrophs can synthesize all essential organic molecules from simple inorganic substances, such as H_2O, CO_2, and NH_3, and some energy source (sunlight, for example). Contemporary examples of autotrophs are algae, plants, and certain bacteria.

The most successful autotrophs were those that evolved a system of utilizing solar energy — *photosynthesis* (Chapter 7). The advent of photosynthesis is probably the second most important event (after the creation of life itself) in the evolution of our present biosphere. Most of the oxygen of our atmosphere is thought to be a by-product of photosynthetic organisms. As oxygen accumulated over hundreds of millions of years, it oxidized the poisonous gases of the ancient atmosphere, and permitted the evolution of aerobic organisms from anaerobic organisms. The term **aerobic** refers to any biological process that can occur in the presence of molecular oxygen. In contrast, **anaerobic** refers to any biological process that cannot occur in its presence.

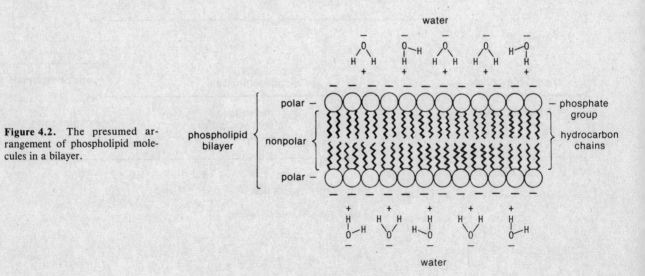

Figure 4.2. The presumed arrangement of phospholipid molecules in a bilayer.

CELL STRUCTURE

The first cells were probably little more than microscopic bags of **protoplasm**, the organized colloidal complex of organic and inorganic molecules that constitutes the living matter of a cell. Protoplasm was surrounded by a **cell membrane**, sometimes called a **plasma membrane**. In modern cells, a cell membrane is 75-100 Å in thickness).[1]

By separating the protoplasm of the cell from its external environment the cell membrane enables the cell to achieve a state of **homeostasis**. Homeostasis is the maintenance of a relatively stable internal physiological environment in an organism. The cell membrane functions homeostatically by controlling the substances that enter (gases, fuel and food items, hormones, water, etc.) and exit (gases, wastes, hormones, water, etc.) the cell. The regulation of these exchanges depends on the physical and chemical properties of cell membranes, and those of the substances that enter and exit the cell (Chapter 5).

Membrane Structure

The contemporary model of cell membrane structure was first proposed by S. J. Singer and G. L. Nicolson in 1972. The **fluid-mosaic model**, as it is called, describes the cell membrane as a *bilayer of phospholipid molecules* (Figure 4.3). The *hydrophilic* (water-soluble) portions of the phospholipid molecules of each layer collect to form the membrane's outer surface. Each layer's *hydrophobic* hydrocarbon chains project into the center of the membrane where they interact with the hydrophobic portions of the opposite phospholipid layer. The weak intermolecular forces, which hold the bilayer together, allow individual molecules of phospholipid to move relatively freely within each layer. Consequently, cell membranes are highly fluid in nature, and yet have the ordered structure of crystal. Cholesterol molecules are incorporated into the hydrophobic regions of the membrane, and further increase the fluidity of the membrane by disrupting the tight packing otherwise possible in clusters of straight hydrocarbon chains.

Also according to the Singer-Nicolson model, proteins are scattered in the phospholipid bilayer (Figure 4.3). Some of them, the **integral proteins**, extend through the entire thickness of the membrane to be exposed at each surface; others, the **intrinsic** or **peripheral proteins** are only partially embedded, and are exposed at only one surface. Some integral proteins are thought to be carrier molecules (**channel proteins**) involved in the transport of materials across the membrane. Other proteins presumably function as enzymes regulating specific chemical reactions. In both cases the proteins are probably synthesized within the cell, and incorporated into the cell membrane.

On the external surface of the plasma membranes of animal cells, many membrane proteins and some membrane lipids are conjugated with short polysaccharide chains (Figure 4.3). These glycoproteins and glycolipids project from the surface of the bilayer and

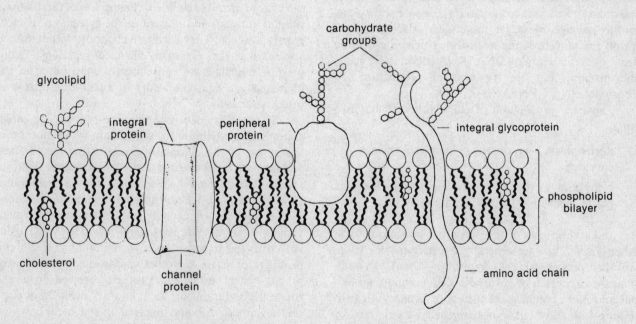

Figure 4.3. Cell membrane structure — the fluid-mosaic model (Singer and Nicolson, 1972).

form an outer coating, termed the **glycocalyx**. It appears to vary in thickness in difference cell types. (It is not known whether an analogous layer exists on internal cell membranes or only on the external surface.) The function of the glycocalyx is obscure. It may simply *provide the mechanical and/or chemical protection for the cell membrane*. However, there is evidence suggesting that it may be involved in *cell recognition phenomena*, in *the formation of intercellular adhesions*, and *in the absorption of molecules through the cell surface*.

Prokaryotic and Eukaryotic Cells

Early cells were simple. Basically, they possessed protoplasm enclosed within a cell membrane (Figure 4.4a). These early cells are referred to as **prokaryotes**. (Bacteria are examples of prokaryotic life today.) About 1.5 billion years ago, however, the **eukaryotes** developed. Eukaryotes differ from the prokaryotes most notably in the organization of their genetic material.

In prokaryotic cells genetic material is in the form of a large, single DNA (deoxyribonucleic acid) molecule, often localized in the **nucleoid** (a central region of the cell). In eukaryotic cells, however, DNA is associated with proteins (**histones**) in complex structures known as **chromatin filaments**, and collectively known as **chromatin** (Figure 4.4b). Chromatin is surrounded by a double membrane, the **nuclear envelope**, thus forming a **nucleus**. A narrow space, the **perinuclear space**, separates the membranes of the nuclear envelope. The nucleus communicates with the rest of the protoplasm, to a limited degree, via **nuclear pores**. Although nuclear pores permit ions and small molecules, such as water and monosaccharides to pass freely, they are selective as to the passage of larger molecules. Also contained within the nucleus is the **nucleolus**. Nucleoli are small dense bodies containing DNA, RNA (ribonucleic acid), and protein, and are the sites of production for ribosomal RNA (rRNA).

Thus, the protoplasm of eukaryotic cells has two parts:

nucleoplasm — everything within the nuclear envelope

cytoplasm — everything outside of the nucleoplasm but within the cell membrane

Eukaryotic cells differ from prokaryotic cells in other ways. They are generally larger (typically 10–20 μm) than prokaryotic cells (typically 1–2 μm).[2] In addition, the *organelles* of eukaryotic cells are more numerous and more complex than those of prokaryotic cells (Figure 4.4). Most eukaryotic organelles are defined by systems of enclosing membranes.

Eukaryotic Organelles

Organelles are macromolecular cytoplasmic structures whose molecular structures enhance specific biochemical activities. *The number and shape of organelles in a cell is directly correlated to the function of the cell.*

Mitochondria. Mitochondria (singular, *mitochondrion*) are membrane-bound, generally oval-shaped structures located throughout the cytoplasm of all eukaryotic cells (Figures 4.4b and 4.5). Mitochondria transform oxidized fuels into usable energy. Within the mitochondria energy-rich molecules of ATP (adenosine triphosphate) are synthesized from the energy released when monosaccharides (e.g., glucose) and other substances are enzymatically oxidized (*oxidative respiration*, Chapter 6). Mitochondria are the cell's powerhouse — they provide the primary source of energy for biological work, and are therefore usually found in areas of the cell showing the greatest metabolic activity, and are more numerous in more active cells.

Mitochondria possess an outer membrane surrounding an inner one. The inner membrane is involuted with a subsequent increase in surface area. The resulting folds are termed **cristae** (singular, *crista*). Enzymes involved in the oxidation of fuels are bound to the inner membrane. Thus, the greater the inner surface area of a mitochondrion, the greater its increased capacity to carry on oxidative respiration.

Endoplasmic Reticulum. Endoplasmic reticula are systems of membrane-lined channels and sacculations running through and present in the cytoplasm of most eukaryotic cells. They facilitate transport within the cell and serve as sites for many chemical reactions. Endoplasmic reticula have consequently contributed to the increased size and complexity of eukaryotic cells over prokaryotic cells.

The two types of endoplasmic reticula are **rough endoplasmic reticulum (rough ER)**, in which ribosomes are bound to the cytoplasmic side of the membranes; and **smooth endoplasmic reticulum (smooth ER)**, in which no ribosomes are attached to the membranes (Figure 4.4b). Rough endoplasmic reticulum functions in the synthesis of protein and its transport through vesicles. Smooth endoplasmic reticulum is primarily involved in the transport of raw materials and the end products of protein synthesis. Portions of this membranal system transform proteins received from the rough ER and transport them to other sites. Other parts of the smooth ER are involved in the breakdown of glycogen and fats and the synthesis of lipids.

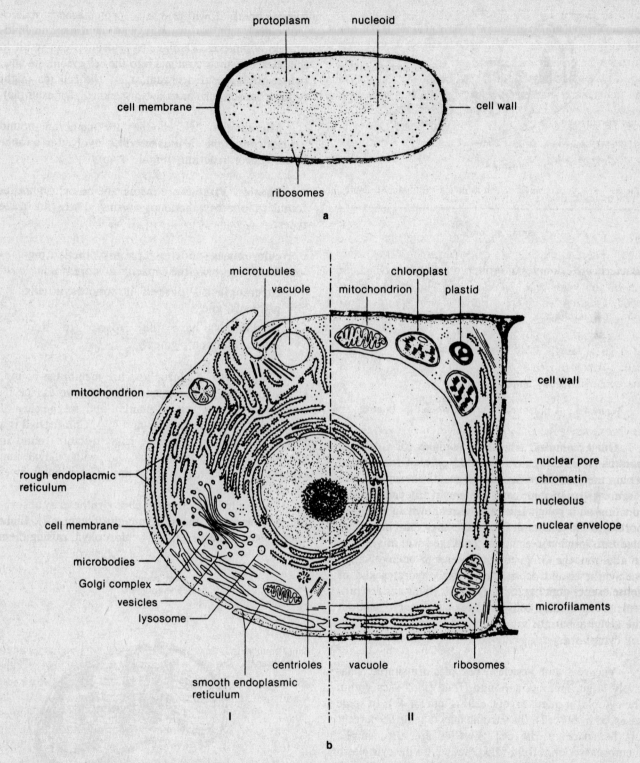

Figure 4.4. A structural comparison of prokaryotic (a) and eukaryotic cells (b). Part b is a comparison of (I) animal and (II) plant cells.

Ribosomes. **Ribosomes** are small, two-part structures (Figure 4.6), which are the sites of protein synthesis in all cell types, prokaryotic as well as eukaryotic. The ribosomes of eukaryotic cells, however, are somewhat larger than those of prokaryotes. These structures are either attached to the endoplasmic reticulum (forming the rough endoplasmic reticulum), or are freely suspended in the cytoplasm.

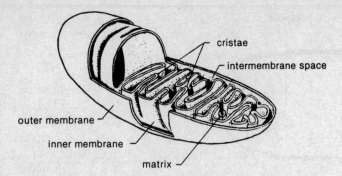

Figure 4.5. A mitochondrion and its double membrane structure.

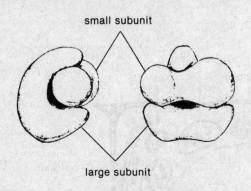

Figure 4.6. A current model of the ribosome (two views).

Golgi Complex. The **Golgi complex** (or **Golgi apparatus**) is a cluster of flattened parallel sacs found within the cytoplasm of many types of eukaryotic cells (Figure 4.4b). In some cases proteins synthesized in conjunction with rough ER have been shown to be transported to Golgi complexes, where they may be assembled into larger molecules or packaged within vesicles. In addition, the Golgi complex seems to be involved in the synthesis and packaging of carbohydrates and of complexes containing carbohydrates, lipids and/or proteins. In specialized cells that have a secretory function, the Golgi apparatus may make up to two-thirds of the cell cytoplasm.

Vacuoles and Vesicles. **Vacuoles** are simple, relatively large, membrane-bound, fluid-filled sacs within the cytoplasm of eukaryotic cells (Figure 4.4b). In some cases, as in fat cells, the vacuole may take up 95 percent of the space of the cell. **Vesicles** are also simple, membrane-bound, fluid-filled sacs within the cytoplasm of eukaryotic cells (Figure 4.4b), but they are relatively small. Both structures function in the storage and transport of substances.

Lysosomes. **Lysosomes** are specialized vesicles (Figure 4.4b), which function in the storage of digestive enzymes. **Lysozymes**, the enzymes contained in lyso-

somes, break down primarily proteinaceous macromolecules, such as proteins, carbohydrate- or lipid-protein complexes. Damage to lysosomes can result in the release of their contents into the cell cytoplasm and, hence, the general degradation of the cell (as might result from sunburn or radiation sickness, for example).

Microbodies. **Microbodies** are membrane-bound vesicles containing various enzymes involved in a range of conversion reactions (Figure 4.4b).

Plastids. **Plastids** are membrane-bound organelles found in photosynthetic organisms. There are three types:

chloroplasts – possess photosynthetic pigments and have the capacity to store starch

chromoplasts – possess nonphotosynthetic pigments, and

amyloplasts – have the capacity to store starch, but lack pigments

Chloroplasts. **Chloroplasts** are membrane-bound, chlorophyll containing organelles in eukaryotes (specifically, algae and terrestrial plants), and are the sites of photosynthesis (Figures 4.4b and 4.7). **Chlorophyll** is a molecule based on the same ring structure found in components of hemoglobin, but with magnesium replacing the central iron atom. (Chlorophyll gives plants their green color.)

Chlorophyll functions in photosynthesis by absorbing specific wavelengths of sunlight (Chapter 7). Light energy excites the electrons of chlorophyll, raising them

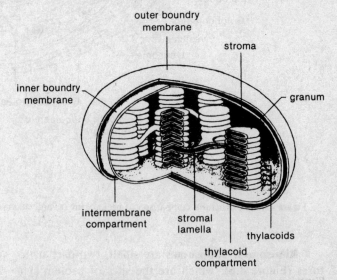

Figure 4.7. The membranous structures and internal regions of the chloroplast.

to higher energy levels. As the electrons return to their original energy levels, through a series of acceptor molecules, the energy released is stored in other molecules, and is used later to reduce carbon dioxide to carbohydrate:

$$CO_2 + H_2O \xrightarrow[\text{chlorophyll}]{\text{light}} (CH_2O)_n + O_2$$

Like mitochondria, chloroplasts are bounded by an external membrane, and contain an internal membrane complex which form structures called grana, composed of stacks of thylakoids (Figure 4.7). Imbedded in those membranes are molecules, such as chlorophyll, enzymes, and coenzymes, which function in photosynthe-sis. The expanded surface area of its inner membrane increases the capacity of the chloroplast for photosynthetic activity.

Microfilaments and Microtubules. **Microfilaments** are extremely fine, thread-like structures composed of the contractile proteins actin and myosin. **Microtubules** are hollow, tube-like structures made mostly of protein (tubulin) subunits (Figures 4.4b and 4.8). Both microfilaments and microtubules contribute to the **cytoplasmic lattice**, an irregular, three-dimensional network pervading and organizing the cytoplasm. In addition, both are involved in various forms of cell movement. For instance, microfilaments are involved in

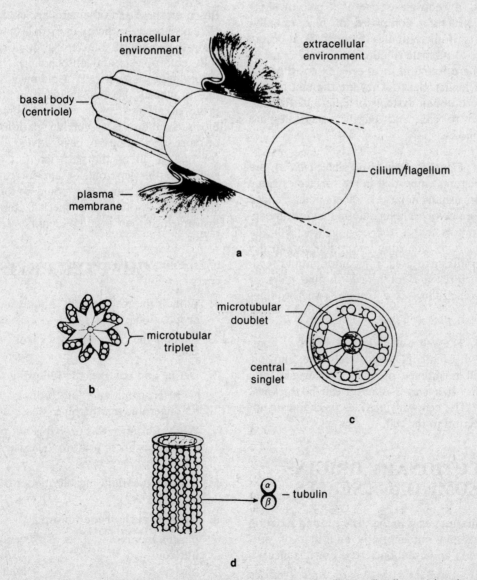

Figure 4.8. Comparative structure of the basal body (centriole) and the axial complex of a cilium or flagellum: (a) junction of basal body and cilium or flagellum at the plasma membrane, (b) cross section of basal body (centriole), (c) cross section of cilium/flagellum revealing structure of axial complex, and (d) microtubule and tubulin subunit.

- the pinching in of a cell (*cytokinesis*)
- a constant directional motion of organelles (*cytoplasmic streaming*) in certain types of plant cells
- the crawling motion of cells across a surface

Microtubules are involved in

- movement of organelles
- the determination of cell shape
- the segregation and distribution of chromosomes during cell division
- the motion of cilia and flagella

Centrioles. A centriole is one of a pair of short barrel-shaped structures comprised of nine radially-arranged triplets of microtubules (Figure 4.8). It occurs at a right angle to its mate (Figure 4.4b).

Centrioles are found in most cells except those of many kinds of higher plants. They are thought to give rise to the microtubular systems of cilia and flagella. When attached to cilia and flagella, centrioles are termed **basal bodies**.

Cilia and Flagella. Cilia (singular, *cilium*) are short, thin structures embedded in the surface of some eukaryotic cells, usually in large numbers and arranged in rows. They are involved in locomotion and the movement of substances across the cell surface. **Flagella** (singular, *flagellum*) are long, whip-like organelles found in eukaryotes and used in locomotion and feeding. Both structures are composed of nine pairs of microtubules encircling two central microtubules (Figure 4.8).

Cell Wall. The **cell wall** is a rigid structure composed of nonliving material, such as cellulose, and surrounds the cell membrane of some prokaryotic and eukaryotic cells. It averages about 1 μm in thickness (Figure 4.4b).[3] The cell wall provides mechanical support and protection to the cell.

EVOLUTIONARY ORIGINS OF SOME ORGANELLES

Many similarities exist in the structure and function of chloroplasts and mitochondria and that of prokaryotes (bacteria and cyanobacteria). For instance:

- Chloroplast and mitochondria are approximately the same size as prokaryotic cells.

- Both possess their own DNA, which is similar in size, shape and structure as that of prokaryotes, and like prokaryotes, is not membrane-bound.
- The structure and shape of their ribosomes are similar to those of prokaryotes.
- The internal membrane structure of chloroplasts bears a striking resemblance to the arrangement of photosynthetic membranes in cyanobacteria.
- Considerable evidence suggests that mitochondrial replication is under cytoplasmic control rather than under control of the nucleus, thus suggesting that mitochondria may, to a certain extent, be genetically independent of the nucleus.

Based on these and other similarities, Lynn Margulis in 1972 proposed the theory of cytoplasmic inclusion to explain the origin of some eukaryotic organelles, specifically chloroplasts and mitochondria.[4] Essentially, the theory suggests that over a billion years ago bacteria-like organisms became included in the cytoplasm of proto-eukaryotes, as a result of some form of symbiosis. **Symbiosis** is a recurring relationship between two or more species that can be positive, negative, or neutral for one or all of the participants (symbionts). With evolution, the symbionts became increasingly dependent upon each other until they could no longer live without each other. As a result of specialization, the symbionts lost much of their original identity.

CHAPTER REVIEW

1. What is the "cell concept"? State and explain its four propositions.

2. Explain the contemporary theory of the origin of cells.

3. Define and contrast the following terms:
   a. heterotroph and autotroph
   b. anaerobic and aerobic

4. What substance was probably not present in the ancient atmosphere prior to the existence of photosynthesis?

5. Explain the fluid-mosaic model of cell membrane structure.

6. Describe each of the following:

plasma membrane	vesicle
nucleoid	lysosome
nucleus	microbody
nuclear envelope	chloroplast
nuclear pore	chromoplast

nucleolus
chromatin
chromatin filament
ribosome
rough endoplasmic reticulum
smooth endoplasmic
  reticulum
mitochondria
Golgi complex
vacuole

amyloplast
microfilament
microtubule
cytoplasmic lattice
centriole/basal body
cilium
flagellum
cell wall

7. What evidence exists to support the cytoplasmic inclusion theory of the origin of eukaryotic organelles, such as mitochondria and chloroplasts?

[1] Refer to Appendix A, "Microscopic Measurements."

[2] *Ibid.*

[3] *Ibid.*

[4] Lynn Margulis, *Origins of Eukaryotic Cells* (New Haven, Connecticut: Yale University Press, 1970).

CELL TRANSPORT MECHANISMS

Cells or cellular systems essentially live in two complex environments: an *external environment* that may be highly variable, and a *homeostatic internal fluid environment*. A biological system lives only as long as biochemically-necessary molecules are taken into the cell, and metabolic waste and nonessential molecules are removed. Unicellular organisms and the individual cells and tissues of complex organisms must maintain a physiologically-guided steady state (homeostasis) or cellular death will occur. Thus, the movement of materials into and out of cellular systems influences the ability of cells to maintain a high degree of metabolic efficiency.

The cell membrane regulates the passage of materials into and out of the cell, a function that makes it possible for the cell to maintain its structural and functional integrity. This regulation depends on interaction between the cell membrane and the materials that pass into or out of the cell. There are two general mechanisms by which substances pass into or out of a cell:

passive transport — the transport of materials across the cell membrane *without* the expenditure of energy on the part of the cell

active transport — the transport of materials across the cell membrane *with* the expenditure of energy by the cell

PASSIVE TRANSPORT

Diffusion and Permeability

Diffusion is the process by which substances passively enter or leave a cell. It is the movement of materials *with* a **concentration gradient**, that is, from an area of higher concentration to one of lower concentration. For instance, sugar molecules of a sugar cube dropped in a cup of water disperse and become evenly distributed throughout the solution. (A **solvent** is a substance, usually liquid, capable of dissolving another substance — the **solute**. The homogeneous mixture of solvent and solute is termed a **solution**.) Diffusion is driven by the kinetic energy inherent in all particles, as manifested by the collision of solvent molecules (e.g., water) with solute particles (e.g., sugar). Consequently, substances diffuse from areas of *higher* kinetic energy to those of *lower* kinetic energy.

The rate of diffusion is influenced by several factors. It is directly related to temperature because thermal energy is converted into the kinetic energy of dissolved particles. It is inversely related to particle size because, for a given quantity of energy, small particles move faster than large particles. In addition, the greater the concentration differential between two regions, the greater the rate of diffusion (because particles will move from an area of higher concentration to one of lower concentration). Finally, the structure, charge, and polarity of molecules may also determine their capacity to diffuse across membranes.

Membranes through which substances can pass are said to be **permeable** to those substances. Nearly all plasma membranes are permeable to water. If water (or some other solvent) is the only substance that can pass through the membrane, the membrane is said to be **semipermeable**. Membranes that display a gradation of permeability — that is, permit water and small ions to pass through more readily than salts or sugars, for example — are said to be **selectively (differentially) permeable**. The permeability of a membrane is determined by the physical relationships between the membrane and substances on either side of it.

When diffusion across a membrane results in equal concentrations of a substance on either side of a membrane, diffusion continues but with no net movement of particles in either direction. Consequently, the regions on either side of the membrane are said to be in **dynamic equilibrium**.

Forms of Passive Transport

Simple Diffusion. Simple diffusion is the movement of small solute particles (ions, gases, and so on) across cellular membranes. The particles become temporarily dissolved in the lipid bilayer along with millions of other particles of various substances. Within a fraction of a second the particles leave the membrane, continuing along their individual concentration gradients. Simple diffusion is the primary means by which materials enter and leave cells, and is also an important mechanism of intracellular transport.

Osmosis. Osmosis is the diffusion of water across a differentially permeable membrane in response to a solute concentration gradient, a pressure gradient, or both. For osmosis to occur, some solute particles must be too large to diffuse across a differentially permeable membrane separating two solutions. The solute concentration of a solution determines the solvent concentration. A solution that has a relatively high solute (low water) concentration is termed **hypertonic**; one which has a relatively low solute (high water) concentration is termed **hypotonic**. Consequently, if a membrane is impermeable to a solute, *the water will diffuse from the area of hypotonicity to that of hypertonicity*. The force generated by water molecules flowing across a membrane is termed **osmotic pressure**. Water molecules will continue to diffuse even after the solute and water concentrations are in dynamic equilibrium — equivalent on both sides of the membrane. However, there will be no further *net* movement of water.

The effect of osmosis on cells is demonstrated by red blood cells (Figure 5.1). Plasma, the fluid portion of blood, is normally **isotonic** to that of the red blood cells. Isotonicity is the condition in which two or more solutions have equal numbers of dissolved particles and therefore the same **osmotic potential**. If the plasma is diluted with water, its solute concentration is decreased, and it becomes hypotonic relative to the cytoplasm of the red blood cells. Water then diffuses into red blood cells from hypotonic plasma, causing the cells to swell.

Just how much water will enter a cell depends on the cell and the degree of hypotonicity of the surrounding medium. Obviously, cells can tolerate only a certain amount of swelling before their membranes **lyse** (rupture), spilling their contents into the surrounding medium. Plant cells do not normally succumb to **osmotic lysis** because cell swelling is limited by the rather inflexible cellulose cell wall. The osmotic pressure contained by the cell wall is termed **turgor pressure**, and contributes to the rigidity of plant tissues. **Crenation** — cell shrinkage — results when animal cells are subjected to hypertonic environments. When plant cells are placed in hypertonic media, they undergo **plasmolysis**; that is, water diffuses from cytoplasmic vacuoles into the space between the cell wall and the cell membrane.

In certain specialized plant cells involved in the transport of organic molecules through the plant tissues, osmosis occurs very rapidly in response to a pressure gradient. This phenomenon is termed **bulk flow**.

Facilitated (Mediated) Diffusion

A variety of lipid-insoluble compounds (monosaccharides and amino acids) can pass through the cell membrane and into the cell at a higher rate than would be expected on the basis of their size, charge, distribution coefficient, or magnitude of concentration gradient. The increased rate of diffusion through the membrane is believed to be facilitated by specific channel-forming membrane proteins (enzymes). This form of

Figure 5.1. Effects of osmosis on red blood cells in plasma of three different degrees of tonicity.

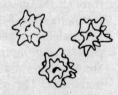

hypertonic   isotonic   hypotonic

passive transport is termed **facilitated** (or mediated) **diffusion**.

The present model of facilitated diffusion suggests that the membrane proteins involved occupy fixed positions, and are held in place by polar and nonpolar surface groups that anchor them to the lipid bilayer. An active site exists at the mouth of a polar channel extending through the membrane (Figure 5.2). This active site seems to be specific to certain molecules or groups of closely related molecules. The collision of an appropriate solute particle with the active site results in the formation of **carrier-solute complex**. This event in turn induces a change in the structure of the active site, resulting in its movement (and that of the solute particle) through the polar channel. In the process the active site's altered structure reduces its affinity for the solute

particle. Consequently, the solute particle is released on the other side of the protein, *Note:* The energy driving this transport mechanism is the kinetic energy of solute particle imparted to carrier proteins upon collision with their active site.

ACTIVE TRANSPORT

In contrast to passive transport, **active transport** is the movement of materials into or out of a cell, *against* a concentration gradient, with the expenditure of energy on the part of the cell. There are principally two forms of active transport: membrane active transport and bulk transport.

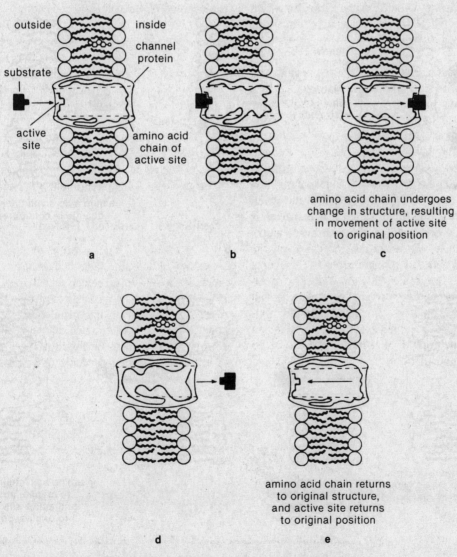

Figure 5.2. Contemporary model of facilitated diffusion.

Membrane Active Transport

During passive transport (diffusion, osmosis, and facilitated diffusion) substances pass through the cell membrane until some sort of equilibrium is achieved. This form of cell transport hinges on the presence of a concentration gradient, with the diffusing substances moving in the direction of the gradient. Substances can also move through the plasma membrane, into or out of a cell, against a concentration gradient. This requires an expenditure of energy on the part of the cell and is called **membrane active transport.**

Channel proteins are also involved in membrane active transport. As with facilitated transport these proteins act on specific substances or those of closely related groups. However, in membrane active transport, cellular energy (resulting from *ATP hydrolysis*) is required to activate the transport process (Figure 5.3).

Bulk Transport

Bulk transport is the process by which large (sometimes macroscopic) portions of liquids and or solids are moved into or out of cells. In bulk transport the cell membrane is still involved, but materials are not transported "through" it.

Bulk transport can be classified by the *direction of movement* or the *mechanism of movement.*

Direction of Movement. The direction (in or out) in which materials are moved determines two fundamental forms of bulk transport:

endocytosis — the formation of cytoplasmic vesicles or vacuoles from the plasma membrane, and the consequent entrapment, within

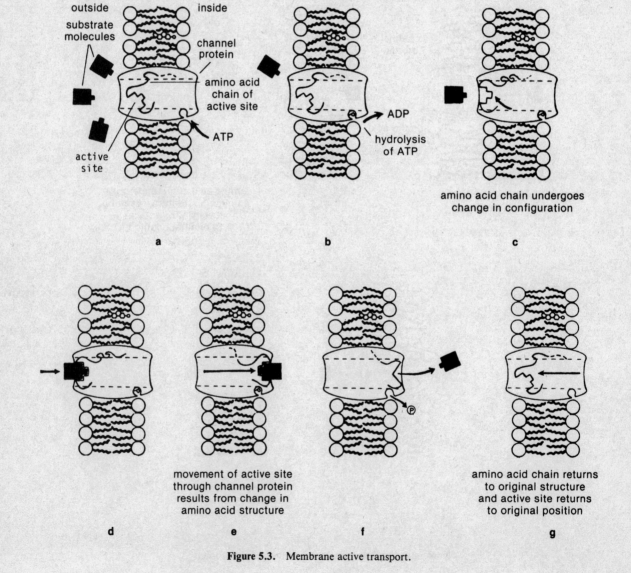

Figure 5.3. Membrane active transport.

these structures, of materials formerly in the cell surroundings

exocytosis — the formation of cytoplasmic vesicles or vacuoles in the packaging of secretory materials or waste products that are transported to the cell surface; there, the membrane of the vacuole fuses with the cell membrane, expelling the contents of the vesicle or vacuole to the outside

The best understood form of exocytosis is *secretion*. When the secretory vesicle or vacuole touches the plasma membrane, lipids in both membranes are moved aside, making the membranes more fluid. After the vesicle's contents have been discharged to the outside, the vesicle membrane is incorporated into the plasma membrane (Figure 5.4).

Mechanisms of Movement. Mechanisms of movement involve basically three types:

pinocytosis — the engulfment of small quantities of extra cellular material (proteins, amino acids, certain ions, water, etc.) in small cytoplasmic vesicles

phagocytosis — similar to pinocytosis but involves the engulfment of much larger quantities of particulate material

rhophecytosis — the transfer of small quantities of cytoplasm, together with their inclusions, from one cell to another

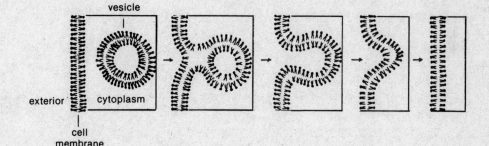

Figure 5.4. The fusion of an exocytic vesicle with the plasma membrane. Endocytosis essentially occurs in reverse (in sequence from right to left).

CHAPTER REVIEW

1. Contrast the terms passive and active transport.
2. Define the following terms:
   diffusion
   concentration gradient
   permeable
   semipermeable
   selectively (differentially) permeable
   crenation
   plasmolysis
   bulk flow
   hypotonicity
   hypertonicity
   isotonicity
   osmotic lysis
   turgor pressure
   dynamic equilibrium
   osmosis
   osmotic pressure
3. Outline and discuss the major forms of passive transport.
4. Compare and contrast membrane active transport and bulk transport.
5. Discuss the various forms of bulk transport, both in terms of direction and mechanism of movement.

6

CELL METABOLISM

The cellular reactions that provide energy for building, maintaining, and degrading cellular components are known as cellular **metabolism**. These reactions use specific starting compounds, are catalyzed by specific enzymes, and yield specific end products. Metabolism proceeds in a stepwise fashion through **intermediary molecules** — that is, the end products of one enzyme-catalyzed reaction become the starting materials for the next reaction. These sequential reactions are known as **intermediary metabolism**. Single-celled *E. Coli* bacteria, for example, carry out over 1000 such chemical reactions. This large number of specific reactions, however, is actually the result of only two metabolic processes — catabolism and anabolism.

Catabolism generates chemical energy; larger molecules are broken down into smaller, simpler molecules, and energy is released. **Anabolism** uses energy to build the larger more complex molecules of the cell. Both occur simultaneously within the cell, yet are independently controlled. The metabolic processes of *most* life forms are very similar, indicating the common origin of cell types.

CELL TYPES AND THEIR ENERGY SOURCES

Cells can be classified on the basis of carbon and energy acquisition. **Autotrophic** cells use CO_2 as the carbon building block from which all other biomolecules are synthesized. These "internally-produced" biomolecules are then used as an energy supply for metabolism. **Heterotrophic** cells break down and rearrange more complex organic sources — sources taken from outside the cell. (Autotrophic organisms include photosynthetic organisms and certain bacteria. Most microorganisms and the higher animals are heterotrophs.)

Heterotrophs are further subdivided on the basis of the final electron acceptor molecule. **Aerobes** use molecular oxygen, O_2, as the final acceptor of the electrons gained from the organic electron donor. Aerobes therefore must live in the presence of oxygen. **Anaerobes**, on the other hand, function in the absence of oxygen, using other molecules as the acceptor of electrons. **Facultative anaerobes** (facultatives) can function aerobically *or* anaerobically. Oxygen is utilized when present, but other organic molecules function as electron acceptors in the absence of oxygen. **Strict anaerobes** cannot use oxygen and may even be poisoned by it.

Cellular energy sources are varied. Cells that trap light energy are **phototrophs**. Oxidation and reduction cycles provide the energy source for **chemotrophs**. Further subdivision of these cell types is based on the electron donor molecule that is oxidized. **Chemoorganotrophs** and **photoorganotrophs** use complex organic molecules, such as glucose, as an electron source. **Chemolithotrophs** and **photolithotrophs** use simple inorganic substances — hydrogen, hydrogen sulfide, ammonia, sulfur, or carbon dioxide — as electron donors. The majority of organisms are chemoorganotrophs or photolithotrophs.

THE ATP CYCLE — ENERGY EXCHANGE IN LIVING SYSTEMS

A constant energy supply is required for the three major functions in living cells:

1. mechanical work (the contraction of muscle cells or impulse transmission of nerve cells, for example)

2. active and facilitated transport of molecules and ions across the cell membrane

3. biosynthesis of macromolecules and cell components

The molecule of energy exchange within the cell is **adenosine triphosphate**, or **ATP** (Figure 6.1). ATP is used by all cells as an energy transfer molecule. It is composed of adenine, a nitrogenous base covalently bonded to a ribose sugar molecule, which in turn is covalently bonded to three phosphates. Energy is stored in the two phosphate–phosphate bonds and released when the bonds are broken. Removal of one phosphate group from ATP results in **adenosine diphosphate**, or **ADP**.

The conversions of ATP to ADP and of ADP to **AMP (adenosine monophosphate)** are *energy-releasing reactions*. These reactions are reversed to rebuild energy-rich molecules. A molecule of inorganic phosphate, P_i, can be covalently bonded to AMP to yield ADP. This process is repeated to form ATP from ADP and P_i. The formation of the covalent bond between adenosine (adenine + ribose) and the phosphate requires an *input* of energy and is known as **phosphorylation**. This cycle of

$$ATP \longrightarrow ADP + P_i + energy / ADP + P_i + energy \longrightarrow ATP$$

sums up the energy exchange system within living cells. Energy-requiring metabolic processes (**anabolic reactions**) derive energy from the breakdown of ATP. Energy-releasing metabolic processes (**catabolic reactions**) store energy in recycled ATP molecules.

ATP PRODUCTION

The production of ATP during aerobic catabolism requires three stages (Figure 6.2). (ATP production from glucose is discussed here even though the cell can derive its energy from a variety of sources.) In Stage I, complex food polymers are broken down into simpler molecules. No ATP energy is generated during Stage I. Stage II involves a variety of reactions specific for each type of molecule formed during Stage I. Fatty acids, glycerol, sugars, and many amino acids are all converted to acetyl CoA. A small amount of ATP is generated during Stage II reactions. Stage II, also referred to as *glycolysis*, is discussed in the section that follows. Stage III consists of the *citric acid cycle* and *oxidative phosphorylation* mechanisms. Most ATP production takes place during Stage III reactions. Anaerobic organisms generate energy through simpler modifications of this scheme.

Glycolysis

Glycolysis is a sequence of enzyme-catalyzed reactions that converts a 6-carbon sugar to a 3-carbon molecule of *pyruvate* while storing energy in ATP. Figure 6.3 illustrates these reactions in a series of steps. Note that the 6-carbon molecule is rearranged and chemically modified only slightly at each step. This allows the energy released at discrete points to be efficiently stored as ATP.

Glucose and other carbohydrates are relatively stable under normal cellular conditions. By activating them with energy, however, they become more reactive. Before any energy is released from the glucose molecule,

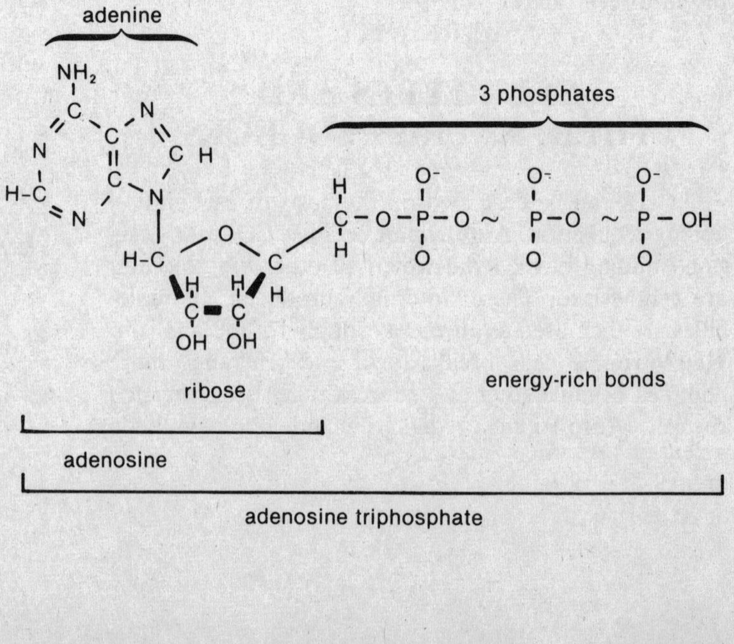

Figure 6.1. Structure of ATP.

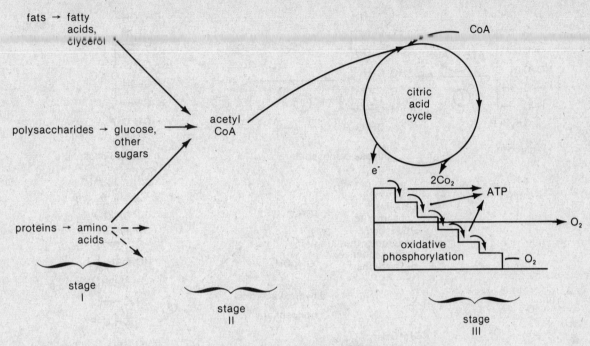

Figure 6.2. The three stages of metabolism (aerobic catabolism).

energy must be added in the form of the transfer of a high energy phosphate group (Step 1). The transfer of a phosphate group is known as **substrate-level phosphorylation**. After another phosphorylation event (Step 3), the 6-carbon fructose di-phosphate molecule is split into two 3-carbon molecules (Step 4) producing dihydroxyacetone phosphate, DHAP, and phosphoglyceraldehyde, PGAL, each with a high energy phosphate group. DHAP is enzymatically converted into PGAL. The reactions shown for PGAL in Figure 6.3 are repeated once for the PGAL molecule converted from DHAP. Up to this point two ATP energy bonds have been used but no energy has been generated.

In the presence of the coenzyme, *nicotinamide adenine dinucleotide, NAD⁺*, electrons are removed from the PGAL molecule, and an additional phosphate group is added (Step 5). A di-phosphate molecule, 3-phosphoglycerol phosphate, is formed. A series of rearrangements and the subsequent release of the phosphate groups (Steps 7, 8, and 9) result in the formation of two ATP molecules as each PGAL is converted to pyruvate.

The amount of energy stored in pyruvate does not differ greatly from the amount stored in glucose. Only a small amount of the total energy in glucose is released during the conversion to pyruvate. About seven percent of the total energy available in the glucose molecule is released and recovered during glycolysis. However, more ATP is generated than consumed: two ATP are used for the initial priming of the 6-carbon molecule;

four ATP are produced. There is a net gain of two ATP molecules for each glucose molecule broken down.

Pyruvate can be broken down in three ways: *fermentation, glycolysis*, and *respiration*. Fermentation in yeast and glycolysis in cells of higher organisms rid cells of wastes, but do not produce additional energy (ATP). On the other hand, under aerobic conditions pyruvate is broken down during respiration and considerable energy is generated.

Anaerobic Fermentation

Fermentation, the conversion of pyruvate to ethyl alcohol, occurs in two steps as shown in Figure 6.4. Carbon dioxide is removed from the pyruvate molecule and the acetaldehyde is oxidized by NADH. This results in the regeneration of NAD⁺ used during glycolysis. The ethyl alcohol produced is transported out of the cell. No additional energy is produced during fermentation reactions. (The wine industry takes advantage of the ethyl alcohol produced by yeast; the baking industry utilizes the carbon dioxide gas produced.)

Lactate Production in Cells of Higher Organisms

The formation of **lactate** from pyruvate occurs in a variety of higher organisms. In humans lactate is produced in the contractile cells of muscle tissue during stress, in red blood cells, and in some cancer cells, for example.

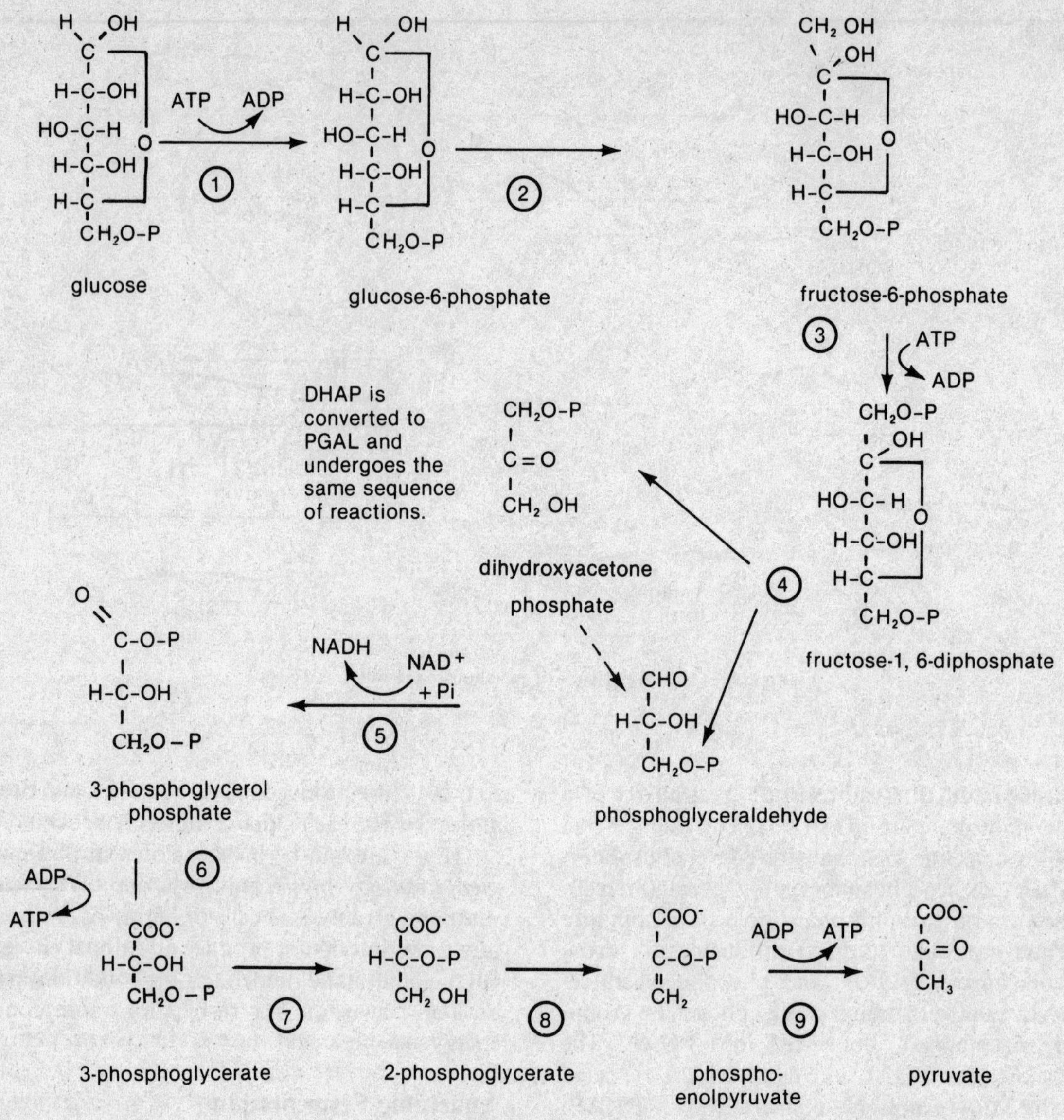

Figure 6.3. Glycolysis.

The normal supply of energy for contractile cells is the ATP generated by the oxidation of glucose. Oxygen in the blood is the oxidizing agent. When considerable amounts of energy are needed and oxygen demand is greater than supply, contractile cells produce ATP by the glycolysis of glucose. (ATP generation during glycolysis can continue even in the absence of oxygen.) Glucose is converted to pyruvate via the glycolytic pathways already discussed. Pyruvate is then reduced to lactate — NADH formed during glycolysis reduces the pyruvate, forms lactate, and regenerates NAD^+ (Figure 6.4). No additional energy is produced.

During intense muscular activity lactate builds up rapidly. The soreness and fatigue experienced after strenuous activity is in part due to increased lactate con-centrations in muscle tissue. This lactate is removed in one of two ways. When the cell's oxygen supply is greater than demand, the reaction described in Figure 6.4 is reversed, and the pyruvate generated enters into the respiration pathway. Alternatively, the lactate can be carried away by the circulatory system to the liver and metabolized back into glucose.

Aerobic Respiration and Oxidative Phosphorylation

Cellular respiration, the process by which organic compounds are completely broken down into carbon dioxide and water, is the most important method of

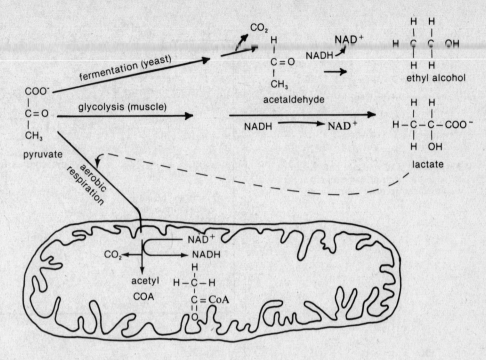

Figure 6.4. Routes of further degradation of pyruvate.

ATP production. It occurs in plants, animals, and most microorganisms. Unlike glycolysis, respiration converts a substantial percentage of the energy originally stored in the organic molecule to ATP. Compared with the two ATP produced during glycolysis, *thirty-eight ATP are produced* per glucose molecule during aerobic cell respiration.

Electrons from oxidized NADH are passed along to molecular oxygen through an electron transport chain — a series of membrane-bound carrier molecules called the **cytochromes**. This results in additional ATP production. The synthesis of ATP, accompanied by the sequential electron flow along the electron transport chain, is known as **oxidative phosphorylation**.

Oxidative phosphorylation takes place within mitochrondria. The glycolysis reactions described earlier take place within the soluble matrix of cytoplasm. The cytoplasmic NADH, produced during glycolysis, is unable to enter mitochondria. A specialized enzyme-shuttle removes electrons from the NADH, and transports them across the inner mitochondrial membrane. Once inside, the electrons reduce the coenzyme flavin adenine dinucleotide, FAD, to FADH₂. FADH₂ enters the electron transport chain at the energy level of coenzyme Q as shown in Figure 6.5.

Each NADH produced during glycolysis can be further reduced *in the presence of molecular oxygen* to yield two additional ATP. One glucose molecule has now yielded two pyruvate and eight ATP; four ATP formed directly during glycolysis, and four ATP formed during oxidative phosphorylation.

The Citric Acid Cycle. The 3-carbon pyruvate molecules are further broken down during cell respiration to release CO_2, H_2O, and additional energy. This process takes place within the mitochondria. The pyruvate is first **decarboxylated** (a CO_2 removed), and joined to coenzyme A (to form acetyl CoA) before entering the complex citric acid cycle (Figure 6.6).

Oxaloacetate, a 4-carbon molecule, is used as a starting material to begin the cycle. Oxaloacetate joins with the two-carbon acetyl CoA, forming a 6-carbon citrate molecule and regenerating coenzyme A. Subsequent rearrangements and decarboxylation result in the formation of the 5-carbon α-ketoglutarate molecule. Loss of electrons to NAD⁺, decarboxylation, and bonding to coenzyme A forms succinyl CoA, a 4-carbon compound. Succinate is then formed by the breaking of an energy-rich bond. This energy is transferred into a molecule of *GTP*, **guanosine triphosphate**. Addition of water and loss of electrons forms oxaloacetate. Another turn of the cycle begins.

Each glucose molecule forms two pyruvate molecules during glycolysis. Entry of these pyruvate molecules into the citric acid cycle yields six NADH, two FADH₂, and two GTP. The hydrolysis of GTP drives the formation of an ATP. Looking back to Figure 6.5, trace the process whereby NADH and FADH₂ produced within the mitochondria generate additional ATP. Each NADH will generate three ATP and each FADH₂ will generate two ATP. The net ATP production from the citric acid cycle is 24 ATP. Eight additional ATP per glucose were formed during glycolysis

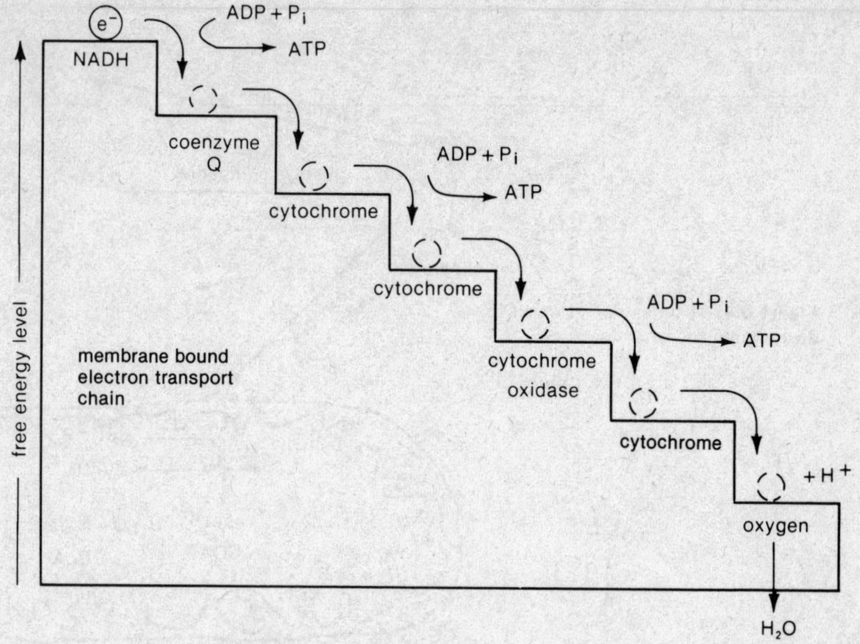

Figure 6.5. The passage of electrons through the electron transport chain with a subsequent decrease in free energy.

and the oxidative phosphorylation of NADH. An additional two NADH are produced within the mitochondria when pyruvate combines with coenzyme A to form acetyl CoA. Since these NADH are formed within the mitochondria, each can drive the generation of three ATP. During aerobic respiration a total of 38 ATP are formed when glucose is converted to carbon dioxide and water (Figure 6.7).

In actuality the efficiency of respiration varies. Laboratory experiments have shown that on the average, 21 to 36 ATP molecules are produced for each glucose molecule that is oxidized. This represents a 25 to 43 percent efficiency of energy recovery from glucose to cellular ATP.

ANABOLISM

Anabolic (biosynthetic) reactions also take place in three stages. Whereas catabolic processes begin with many different polymeric molecules, and converge into a common route of energy production (Figure 6.2), anabolic processes start from a few simple precursor molecules in Stage III and diverge, forming more and varied components at Stage II. Finally, macromolecular cell components are assembled in Stage I. Stage III provides a common route accessible to both anabolic and catabolic pathways, called an **amphibolic pathway**. It serves the dual function of providing small precursor molecules for biosynthesis and of completing Stage II catabolism.

Anabolic synthesis and catabolic degradation of a given precursor to a given product *are not* merely the reverse sequence of reactions. They are often localized within the cell and occur simultaneously. This parallel arrangement of anabolic and catabolic reactions allows for greater cellular control.

REGULATION

The reactions of metabolism are regulated in several ways. One control exists at the **cellular** level. The enzyme systems that catalyze the various reactions are **compartmentalized** — glycolysis takes place within the cytoplasm; the citric acid cycle and oxidative phosphorylation are mitochondrial reactions. The overall rate of a reaction is affected by the rate of exchange of molecules across the mitochondrial membranes.

Individual reaction rates also depend on the *intracellular concentration* of each molecule involved in that reaction: starting material (substrate), end product, cofactors, and enzymes. These concentrations in turn are dependent upon their rate of synthesis *and* degradation.

A third form of regulation involves *regulatory enzymes*. For example, the end product of a sequential reaction may inhibit (turn off) an enzyme involved in an

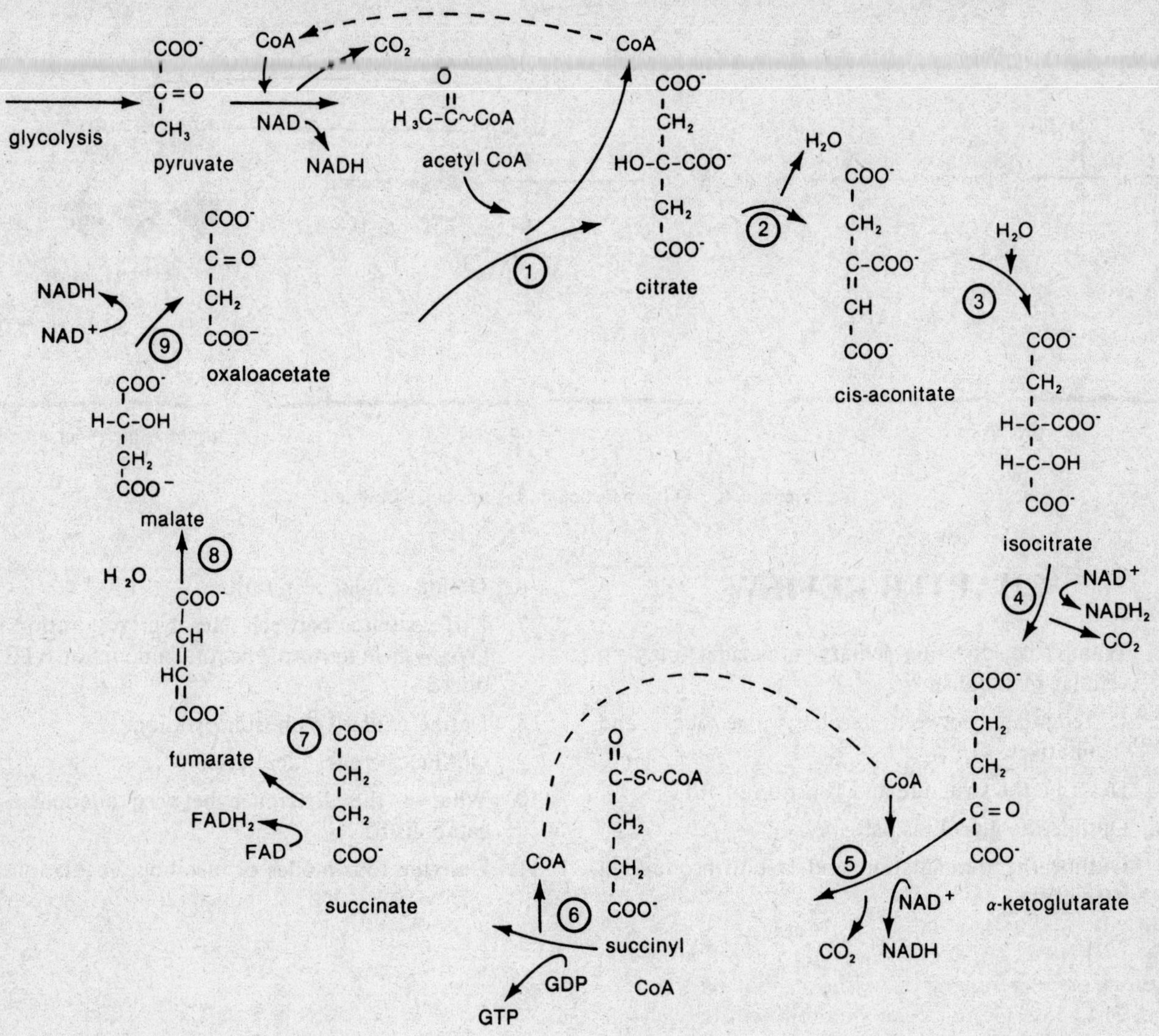

Figure 6.6. Conversion of pyruvate to carbon dioxide and water via the citric acid cycle.

earlier catalytic event. When the end product is plentiful, the sequence of reactions is turned off. As the end product is gradually depleted, its inhibitory influence is decreased and the reactions begin again.

Finally, *genetic regulation* can control the rate of enzyme synthesis. For example, the gene for a particular enzyme may be turned on (actively transcribed) only in the presence of the substrate catalyzed by that enzyme.

In general, the metabolic activity of the cell is controlled by the cell's ATP energy needs. The rate of ATP utilization by the cell corresponds to cellular energy needs at that instant. This close correlation allows cellular flexibility in response to environmental conditions.

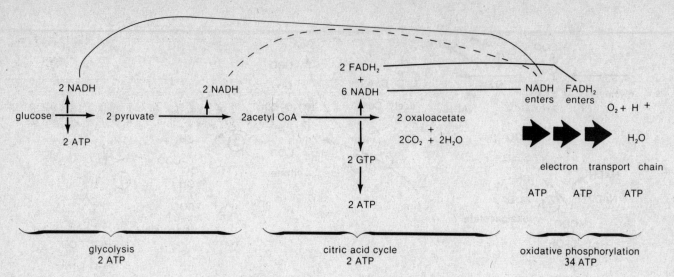

Figure 6.7. ATP production during aerobic respiration.

CHAPTER REVIEW

1. What role do intermediary molecules play in cellular metabolism?

2. Differentiate between aerobes, anaerobes, and facultatives.

3. Describe the structure and function of ATP.

4. Outline the glycolysis pathway.

5. Outline the fermentation and lactate production pathways.

6. Define cellular respiration.

7. Differentiate between aerobic respiration and glycolysis in terms of the total amount of ATP produced.

8. Define oxidative phosphorylation.

9. Outline the citric acid cycle.

10. What is the difference between anabolism and catabolism?

11. Describe four modes of metabolic regulation.

PHOTOSYNTHESIS

The sun is the ultimate source of energy in the biosphere. Solar energy enters the chain of life through organisms that capture and transform the energy during photosynthesis. **Photosynthesis** is the conversion of light energy into chemical bond energy, which is later used to synthesize complex cellular molecules from carbon dioxide and water. In the food chain, photosynthetic organisms are the primary producers of food for other organisms. They *fix* carbon atoms, that is, incorporate energy-poor inorganic carbon into energy-rich organic molecules.

More than 90 percent of all energy used today for heating, lighting, and power is the result of the decomposition of photosynthetic organisms that lived millions of years ago. Photosynthetic organisms of past and present play a vital role in life processes of today.

THE ORIGIN AND EVOLUTIONARY IMPACT OF PHOTOSYNTHESIS

It is generally thought that the first cells of the primordial earth were heterotrophs. Heterotrophic organisms are unable to synthesize their own organic nutrient molecules and must obtain them from an extracellular source. The rich "chemical soup" in which these primitive cells arose provided a source of organic nutrients (Chapter 4), and energy from these nutrients was harnessed by processes of glycolysis (Chapter 6). As the heterotrophs proliferated, presumably for hundreds of millions of years, the organic molecules in their environment became depleted. The concentration of

organic molecules decreased, and carbon dioxide (the end product of glycolysis) in the atmosphere increased. As a consequence, selective pressures favored the survival of organisms able to synthesize organic molecules from abundant inorganic molecules. Those cells that contained genetic information for synthesis of organic molecules were better adapted to their changing environment and became a dominant life form. Cells capable of synthesizing complex organic molecules from inorganic substances are called autotrophs (*auto* = self, *trophos* = one that feeds). However, the combination of simple molecules into complex molecules requires energy — photosynthetic autotrophs use solar radiation as that energy source.

The biosynthetic process of photosynthesis

$$CO_2 + 2H_2O + \text{light energy} \longrightarrow (CH_2O)_n$$

$$+ \ O_2 + \text{chemical bond energy}$$

combines carbon dioxide, water (a source of electrons), and light energy to form a carbohydrate molecule, molecular oxygen, and chemical energy (ATP). Primitive photosynthetic bacteria used electron sources as varied as: hydrogen sulfide, hydrogen gas, thiosulfate, elemental sulfur, ethanol, acetate, and isopropanol. However, the most important photosynthetic autotrophs were those that used water (H_2O) as the electron source. These organisms released free oxygen into the atmosphere as a by-product of photosynthesis, and the increased oxygen concentrations aided in the evolution of aerobic organisms. Accumulated oxygen absorbs the sun's ultraviolet radiation, which can break the hydrogen bonds of organic macromolecules. Once this destructive force was effectively limited, macromolecular structure and subsequent life forms became increasingly complex. Today's life forms continue to depend on photosynthetic autotrophs as an oxygen source *and* a source of carbon containing nutrients.

PHOTOSYNTHETIC STRUCTURES

Photosynthesis takes place in a variety of organisms, both prokaryotic and eukaryotic. The chemical reactions described below are similar in both cell types. However, these reactions are membrane-bound in the photosynthetic prokaryotes; organelle-bound in photosynthetic eukaryotes.

Chloroplasts

In eukaryotes, photosynthesis occurs within specialized cellular organelles called **chloroplasts**. (Figure 7.1 illustrates a photosynthetic organism and the location, relative size, and internal structures of its chloroplasts.) The number of chloroplasts varies from one large chloroplast in *Spirogyra*, to as many as 40 chloroplasts per cell in higher plants. Plant leaves are estimated to have 250 million chloroplasts per square inch. Chloroplasts vary in size from 1 to 10 μm in diameter.

The structure of the double-membraned chloroplast is complex. The inner membrane forms a system of flattened sacs called **thylakoids**. Individual thylakoids are layered, like a stack of pancakes, forming **grana**. The **stroma** is the soluble matrix that surrounds the grana. **Stroma lamellae** connect separate grana.

Chlorophylls and Accessory Pigments

The **chlorophylls** are the organic pigment molecules that capture light energy. These photoreceptor molecules are membrane-bound. Prokaryotic photosynthetic cells contain a variety of chlorophylls; most abundant are the **bacteriochlorophylls**. In eukaryotes, chlorophylls are found in the internal thylakoid membrane. All eukaryotic photosynthetic cells contain chlorophyll a; the higher plants and green algae also contain chlorophyll b; brown algae, diatoms, and dinoflagellates contain chlorophyll c (Figure 7.2).

Chlorophyll gives plants their characteristic green color. **Carotenoids** are accessory pigments that contribute to the yellow, red, and purple colors of some photosynthetic organisms; **phycobilins** are red and blue accessory pigments found in the red algae and the cyanobacteria.

Each photoreceptor molecule has a different absorption maxima, that is, each is able to best utilize light of a particular wavelength. The chlorophylls and accessory pigments complement each other so as to utilize nearly the full range of solar wavelengths that reach the earth. Cells with photoreceptor pigments that capture the most energy are better adapted to their environment.

THE PHOTOSYNTHETIC PATHWAY

Photosynthesis is divided into two parts, the light and dark reactions. The *light reaction* requires the presence of light. Two distinct reactions convert light energy into chemical bond energy. The *dark reaction* does not require light. The energy stored in the cellular energy-carrying molecules ATP and NADPH is used to combine carbon dioxide into monosaccharide molecules.

The Light Reaction

During this first stage of photosynthesis, light energy trapped by chlorophyll and accessory pigments is converted to chemical bond energy and stored in molecules of ATP and NADPH. Chlorophyll, accessory pigments, proteins, lipids, and coenzymes form membrane-bound **photoreaction centers**. These centers, Photosystem I and Photosystem II (P700 and P680, respectively, in some texts), serve as relay antennae, absorbing and transmitting light energy.

Photons (specific quantities of light energy) are absorbed by chlorophyll and accessory pigments within photoreceptor sites. This energy transfer raises two electrons from chlorophyll to a higher energy level (Figure 7.3). These electrons are then transferred to a series of electron carrier molecules, *cytochromes*, in the *electron transport chain*. Chlorophyll oxidizes water (removes electrons from the electron source) to replace the two electrons it transferred. The water molecule is split, producing hydrogen protons and releasing free oxygen.

As electrons are passed along the cytochrome proteins of the electron transport chain, they move to lower energy levels (Figure 7.3). The energy released at discrete points along the chain is stored in a high-energy phosphate bond of ATP.

At the end of the electron transport chain, electrons arrive at **Photosystem I**. (Photosystem I and II were named in the order of their discovery in the laboratory.) A second light reaction boosts the electrons to an energy level even higher than that in Photosystem II. The primary electron acceptor protein of Photosystem I passes the electrons to *ferredoxin* whose iron molecule is alternatively reduced (gains electrons) and oxidized (loses electrons). Ferredoxin passes the electrons to an enzyme that catalyzes the reduction of nicotinamide adenine

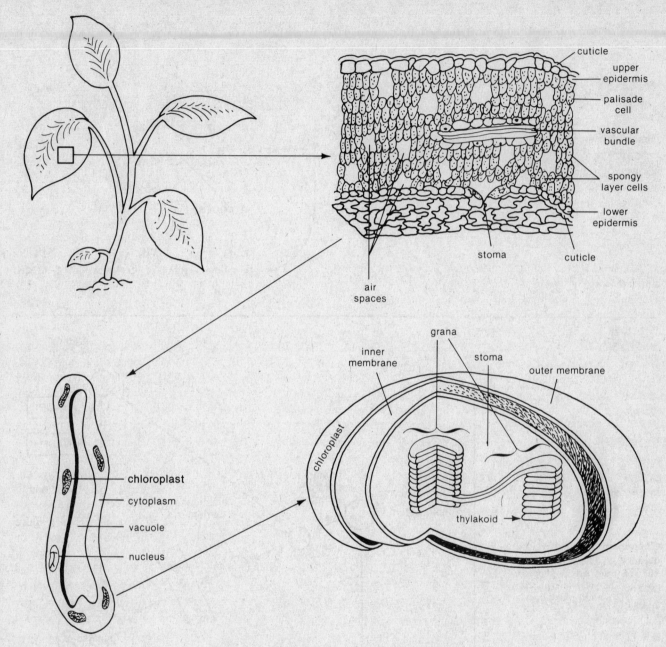

Figure 7.1. Photosynthetic structures of higher plants.

dinucleotide phosphate, NADP⁺. The NADP⁺ molecule accepts the electrons and picks up a hydrogen proton, forming another cellular energy carrying molecule, NADPH.

The energy end product of Photosystem I is NADPH. The energy end product of Photosystem II is ATP. Both are formed as the electrons are passed through the electron transport chain. This unidirectional flow of electrons, from water to the pigments of Photosystem II to the electron transport chain to Photosystem I and eventually to NADPH, is known as **noncyclic photophosphorylation**. However, when the cellular level of NADP⁺ is insufficient, electrons flow from the primary electron acceptor protein of Photosystem I back to the cytochromes and on to *plastocyanin*. One molecule of ATP is generated via this transfer. Electrons can be shuttled through this system of carrier proteins continually during **cyclic photophosphorylation**.

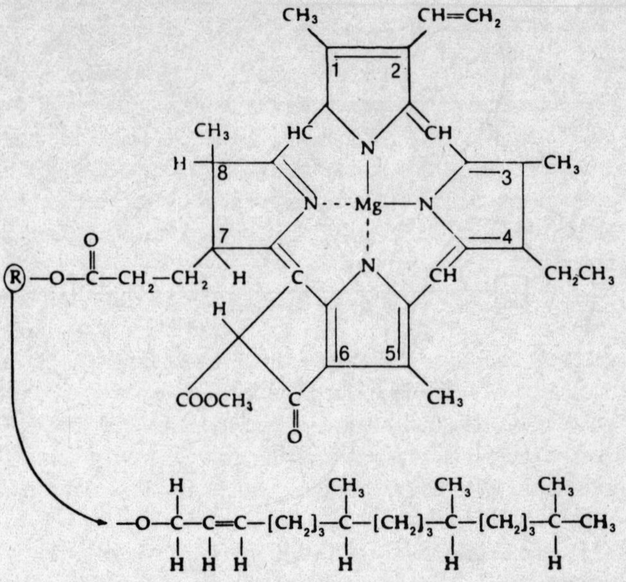

Figure 7.2. The chlorophyll molecule. Chlorophyll a is shown. Chlorophyll b is chemically modified at position 3. Chlorophyll c is modified at position 2. Bacteriochlorophylls are modified at several positions, including 2, 3, 4, 5, and 7.

Figure 7.3. Energy diagram of electron transport in photosynthesis. (The solid line indicates noncyclic photophosphorylation.)

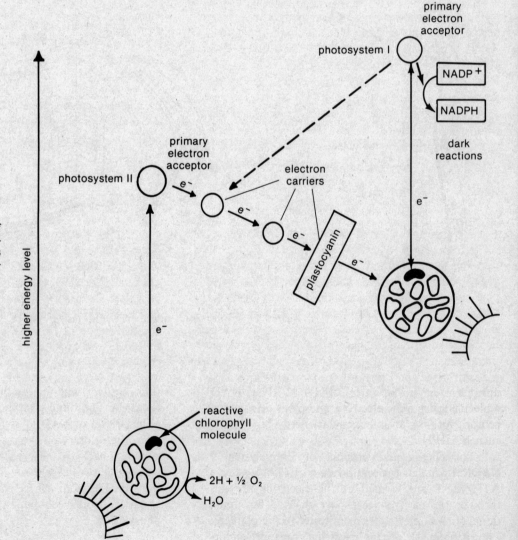

The Dark Reaction

During the dark reaction, plants manufacture monosaccharide sugars from carbon dioxide using the ATP and NADPH molecules produced during the light reaction. In contrast to the light reaction, which is associated with thylakoid membrane-bound units, the dark reaction occurs within the stroma, the soluble matrix of the chloroplast.

The mechanisms of carbohydrate formation fall into two distinct categories. One group of plants, **C_3 plants**, convert carbon dioxide into carbohydrates via a 3-carbon intermediate molecule. The other group of plants, **C_4 plants**, convert carbon dioxide to carbohydrates via a 4-carbon intermediary. C_3 and C_4 plants differ both anatomically and biochemically.

The Calvin-Benson Cycle: The C_3 Pathway. Those plants utilizing the C_3 pathway add carbon dioxide onto an existing 5-carbon sugar — *ribulose diphosphate*, or *RuDP*. RuDP is converted into two 3-carbon molecules of *3-phosphoglycerate*, or *PGA*. PGA is then converted to a 6-carbon sugar molecule, *fructose 6-phosphate*, as energy from two ATP and one NADPH is transferred into the growing carbohydrate molecule. Some of the PGA is built into RuDP, replenishing the molecular building material. One high-energy ATP bond is used in this process. The reaction in C_3 plants can be summarized as follows:

$$6 \text{ RuDP} + 6 \text{ CO}_2 + 18 \text{ ATP} + 12 \text{ NADPH}$$
$$+ 12 \text{ H}^+ + 12 \text{ H}_2\text{O} \longrightarrow 6 \text{ RuDP}$$
$$+ \text{ fructose} + 18 \text{ P}_i + 18 \text{ ADP}$$
$$+ 12 \text{ NADP}^+$$

The Hatch-Stack Cycle: The C_4 Pathway. In C_4 plants, carbon dioxide is added to the 3-carbon molecule *phosphoenolpyruvate*, or *PEP*. A variety of 4-carbon intermediates are formed. This reaction occurs in specialized mesophyll cells. The 4-carbon intermediary is shunted to bundle sheath cells surrounding the plant veins. A molecule of carbon dioxide is removed in a decarboxylation reaction, forming a 3-carbon molecule, pyruvate. The carbon dioxide is passed to the Calvin-Benson pathway and eventually incorporated into fructose-6-phosphate as described earlier. Pyruvate is routed back to the mesophyll cells where it is converted into phosphoenolpyruvate, the starting molecule of the C_4 pathway. This conversion uses two high-energy ATP bonds. The reaction in C_4 plants can be summarized as follows:

$$6 \text{ PEP} + 6 \text{ CO}_2 + 30 \text{ ATP} + 12 \text{ NADPH}$$
$$+ 12 \text{ H}^+ + 24 \text{ H}_2\text{O} \longrightarrow 6 \text{ PEP}$$
$$+ 6 \text{ fructose} + 30 \text{ P}_i + 30 \text{ ADP}$$
$$+ 12 \text{ NADP}^+$$

Comparison of C_3 and C_4 Plants. The net reactions for C_3 and C_4 plants indicate that C_4 plants require more ATP to synthesize one monosaccharide. C_4 plants may seem to be duplicating their effort by fixing carbon dioxide in one cell type and then decarboxylating the molecule and refixing it in another cell type. Despite this seeming contradiction, C_4 plants can: synthesize carbohydrates faster and therefore grow more quickly than C_3 plants, and continue to work efficiently in full sunlight. They require 0–10 ppm (parts per million) of carbon dioxide to continue photosynthesis. This, coupled with mechanisms for reducing water loss, gives C_4 plants a selective advantage in desert and semidesert regions. C_3 plants, on the other hand, require 30–70 ppm carbon dioxide, photosynthesize maximally at one-fourth to one-third full sunlight, and are only moderate carbohydrate producers. C_4 plants, among the most productive plants on earth, include sugarcane, corn, sorghum, crabgrass, and a variety of tropical and desert plants.

PRODUCTS OF PHOTOSYNTHESIS

During the light reaction, solar energy is transferred, converted, and stored in cellular energy molecules. Depending on a cell's metabolic requirements, the chemical bond energy of ATP and NADPH can be used for the entire range of cellular functions. The monosaccharides formed during the dark reaction are the starting molecules for other cellular molecules. Starch and cellulose, two complex carbohydrates characteristic of most photosynthetic organisms, are polymers built from repeating monosaccharide units. The addition of nitrogen to the monosaccharide molecule results in the formation of amino acids and nitrogenous bases, building blocks for proteins and nucleic acids. All of these compounds provide structural components for the photosynthetic organism as well as nutrient sources for other organisms.

CHAPTER REVIEW

1. Define photosynthesis and include the chemical equation.

2. Describe the role of photosynthetic organisms in the evolution of our present-day atmosphere.

3. Describe the structures found within the eukaryotic cell organelle specialized for photosynthesis; give the function of each.

4. Describe the role pigments play in the photosynthesis process.

5. Differentiate between the light and dark reactions.

6. Describe the C_3 and C_4 pathways.

CELL REPRODUCTION

Growth and reproduction are common to all organisms, and both ultimately depend on processes that occur at the cellular level. Growth at the cellular level involves the synthesis of components (proteins, nucleic acids, lipids, and carbohydrates, for example) that contribute to the build up of cell structures. Typically, when a cell has attained a particular size (as defined by its genetic information, the environment, or both), it *divides*. Because the resulting daughter cells originated from one parent cell, this process is regarded as a form of **asexual reproduction**. The hereditary information in the daughter cells is identical to that of the parent cell; thus, the daughter cells are termed *clones*.

For multicellular organisms, growth involves an increase in the number of cells — a consequence of asexual reproduction. In multicellular development, **differentiation**, the process by which like cells become *specialized* (different in form and function), accompanies cell proliferation. Multicellular organisms can reproduce asexually (sporulation, budding, and so on) and sexually. **Sexual reproduction** involves the production of sex cells, or **gametes** (eggs and sperm), from two parents or parent cells. The production of sex cells, **gametogenesis**, is a result of a specialized type of cell division termed **meiotic cell division**. Complementary sex cells join in **fertilization** to form a **zygote** which, in the case of multicellular organisms, undergoes **mitotic cell divisions** to ultimately give rise to the mature organism. The offspring, then, is the product of hereditary information contributed by two parents.

Clearly, an understanding of cell division is essential to understanding mechanisms of reproduction and related biological phenomena (such as inheritance, mutation, differentiation, regeneration, or degenera-

tion). Congruently, cell division must be considered in the context of the life cycle of the cell, just as reproduction must be studied in the context of the life cycle of an organism. A **life cycle** is a recurrent set of events involving reproduction, growth, and development.

ASEXUAL REPRODUCTION

Prokaryotic Fission

Prokaryotic fission is a form of bacterial replication that involves the growth of the cell membrane such that the protoplasm is divided between two daughter cells (Figure 8.1). The process begins in a single cell with the attachment of the DNA molecule to the plasma membrane. The DNA then replicates into two DNA molecules attached to the same membrane. Membrane growth between the two molecules results in their migration and subsequent segregation into the developing daughter cells. Membrane growth and simultaneous cell wall deposition continues across the midsection of the dividing cell, ultimately dividing the cell in two. This form of cell division is termed **binary fission**, because it results in the production of two identical daughter cells, or **clones**. (Eukaryotic cells also have a form of binary fusion.)

Mitotic Cell Division

The cell cycle. The period between the end of one eukaryotic cell division and the end of the next is termed the **cell cycle** (Figure 8.2). It is divided into two sets of

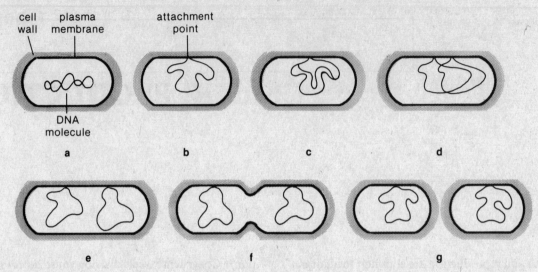

Figure 8.1. Prokaryotic fission: (a) generalized bacterial cell, (b) attachment of DNA to plasma membrane, (c) DNA replication, (d) membrane growth with (e) resulting separation of DNA molecules, (f) beginning of cytoplasmic division (inward growth of plasma membrane and cell wall), and (g) cell division completed.

events: interphase and cell division. **Interphase**, the interval between cell divisions, in turn involves three phases:

G_1 – the period of cell growth prior to DNA replication

S – the period during which DNA replication (synthesis) occurs, resulting in the formation of attached pairs of chromatin filaments

G_2 – the period after cell division during which the cell prepares for cell division

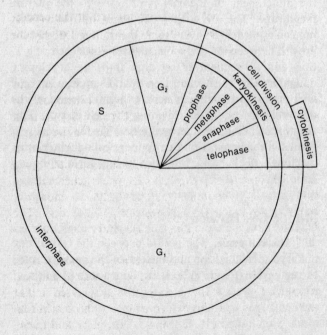

Figure 8.2. Cell cycle.

The events of **cell division** are categorized into two processes:

karyokinesis – the division and segregation of hereditary material to the developing daughter cells

cytokinesis – the division and segregation of cytoplasmic material to the developing daughter cells

Karyokinesis occurs in two forms: mitosis and meiosis. Thus, mitotic cell division is that form of eukaryotic cell division that involves mitosis. It is an integral part of all forms of eukaryotic asexual cell reproduction.

Mitosis. **Mitosis** is the process of nuclear division in which the number of individual DNA molecules is conserved from one cell generation to the next. Mitosis involves a series of events that are categorized into four sequential phases: prophase, metaphase, anaphase and telophase (Figures 8.2 and 8.3).

Prophase begins with the superwinding and superfolding of the replicated chromatin filaments to form **chromosomes**. The process of chromosome formation is termed **condensation**. At this point, each chromosome comprises two **sister chromatids** (Figure 8.4). Each chromatid, in turn, comprises a condensed DNA molecule and associated proteins. The **primary constriction**, where the paired nucleoproteins (nucleic acids complexed with protein) are held together, is termed the **centromere**.

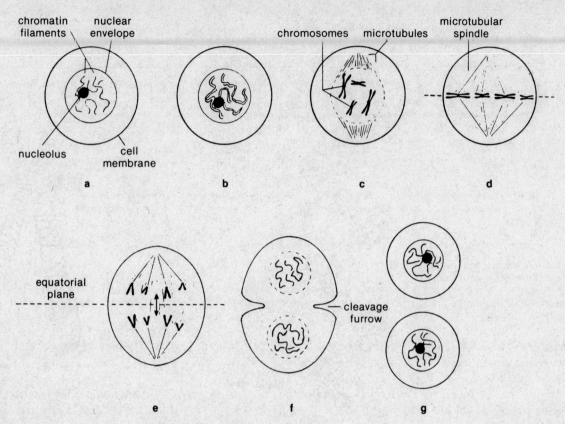

Figure 8.3. Generalized diploid eukaryotic cell in stages of mitotic cell division: (a) G_1 of interphase, (b) G_2 of interphase, (c) prophase, (d) metaphase, (e) anaphase, (f) telophase, and (g) cell division completed.

Simultaneous with condensation: the nucleolus and nuclear envelope seem to disperse and disappear; microtubules appear and become organized to give rise to the **microtubular spindle (spindle apparatus)**. Most microtubular spindles assume one of two shapes: *astral* (starlike) and *anastral* (nonstarlike) (Figure 8.5). In those cells that possess them (animal, protozoan, and lower plant cells), **centrioles** occur in pairs (**centrosomes**) that are replicated prior to cell division. During prophase, the centrosomes migrate to opposite poles of the cell. Microtubular segments organize around the centriole pairs and seem to radiate from them (hence, the term "astral"). In those cells that lack centrioles (in higher plant cells, for example), microtubules form a barrel-shaped spindle lacking the terminal **asters** (hence, the term "anastral").

The beginning of **metaphase** is marked by the absence of the nuclear envelope (Figure 8.3d). This stage of mitosis involves three major processes:

1. The microtubular spindle moves into the region previously occupied by the nucleus.

2. Microtubules of the spindle apparatus become attached to each chromosome at its centromere.

3. The chromosomes migrate to the midpoint or **equatorial plane** of the cell.

Associated with each chromatid located at the centromere is a **kinetochore** (Figure 8.4), the point of attachment for microtubules to the chromosomes.

Anaphase is characterized by the segregation and distribution of sister chromatids of each chromosome to opposite poles of the dividing cell (Figure 8.3e). At least two mechanisms seem to be responsible for the migration of the chromatids. First, microtubules attached to the kinetochores of each chromatid shorten, pulling the sister chromatids apart. Second, other microtubules that extend from pole to pole begin to elongate, resulting in the overall elongation of the cell. The separated sister chromatids subsequently constitute individual "chromosomes".

The arrival of the chromosomes to opposite poles of the dividing cell initiates the last stage of mitosis, **telophase** (Figure 8.3f). Condensation is reversed so that each chromosome is transformed into a chromatin filament. Simultaneously, the nuclear envelope and nucleolus reappear, while the mitochondrial spindle seems to

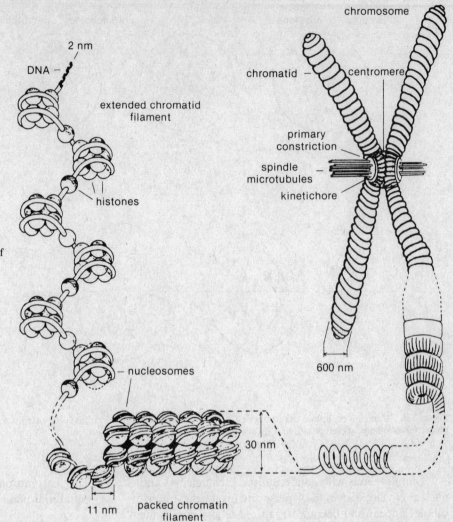

Figure 8.4. A contemporary model of metaphase chromosome structure.

disintegrate. Completion of the reformation of the nucleus marks the termination of this phase (Figure 8.3g).

Cytokinesis. Cytokinesis usually accompanies mitosis in asexual reproduction. Typically, it extends from late anaphase through telophase. In most animal cells, cytokinesis begins with the accumulation of patches of dense, apparently structureless material about the spindle apparatus, in the region of the equatorial plane. A layer of this material forms completely across the dividing cell and is termed the **midbody**. As the midbody develops, a shallow depression termed the **cleavage furrow** develops at the margin of the equatorial plane (Figure 8.3f). The cleavage furrow deepens as microfilaments attached to the plasma membrane pull it inward. Subsequently, cytoplasm is roughly divided be-

tween the developing cells, and the inward movement of the cell membrane ultimately bisects the original cell into two daughter cells.

As with animal cells, the formation of a layer of material on the equatorial plane initiates cytokinesis in higher plant cells. However, these cells have relatively rigid cell walls that are not conducive to cleavage furrow formation. Instead, a **cell plate** is formed that begins with the accumulation of vesicles (derived from the Golgi complex) about the equatorial plane (Figure 8.6). The vesicles contain pectin and various polysaccharides. As vesicles continue to accumulate and fuse, their contents contribute to the formation of the cell plate, a partition between the daughter cells that has the same composition as the cell wall. The membranes of the fused vesicles contribute to the formation of cell membranes lining both sides of the cell plate.

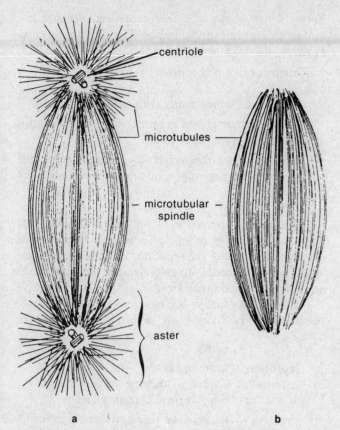

Figure 8.5. Microtubular spindles: (a) astral spindle; (b) anastral spindle.

Other Forms of Asexual Reproduction

Additional forms of asexual reproduction include: budding, sporulation, fragmentation, regeneration, vegetative reproduction and parthenogenesis.

Budding involves the development of a new organism as an outgrowth of the parent organism. In the case of unicellular eukaryotes (e.g., yeasts), the *nucleus divides equally*, but the *cytoplasm divides unequally*. The result is the formation of a large mother cell and small daughter cell. In animals (e.g., *Hydra*), a new individual develops as an outgrowth of the parent body. The cells of the developing bud proliferate and differentiate into a separate set of organs. Ultimately, the new organism breaks away achieving autonomy.

Some prokaryotic and eukaryotic organisms reproduce by producing asexual cells, **spores**, capable of developing into an adult form through mitotic cell division. **Sporulation** is the process of spore formation. In prokaryotes and many eukaryotes, spores enclosed within thick, protective walls serve to increase the survival of the species by being able to survive adverse environmental conditions.

Fragmentation is the process by which branched or unbranched multicellular prokaryotic filaments are reproduced from pieces broken off from the original colony. The cells of the fragments undergo prokaryotic fission, resulting in the growth of new colonies.

Regeneration is the process by which certain cells of multicellular organisms differentiate and proliferate to replace other cells that have been lost or destroyed (as in the healing of small cuts in human beings, for example). Flatworms, earthworms, and starfish are multicellular organisms possessing great capabilities of regeneration.

Vegetative reproduction is a form of asexual plant reproduction. Cells from unspecialized or specialized tissues or organs, such as runners, tubers, or bulbs of plants, proliferate and differentiate resulting in a mature plant.

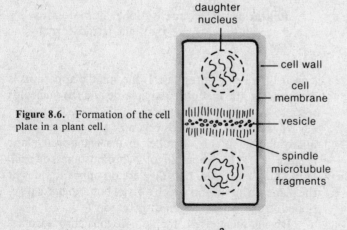

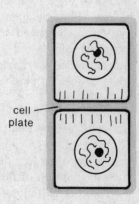

Figure 8.6. Formation of the cell plate in a plant cell.

Parthenogenesis is the development of a multicellular organism from an unfertilized egg. In animals (e.g., wasps and bees), mechanical stimulation seems to trigger mitotic cell division of the egg and its subsequent development into the mature organism. In plants (sunflowers, orange trees, or grape vines, for example), hormones have been identified as the triggering agents.

SEXUAL REPRODUCTION

Conjugation

Although technically *not* a form of sexual reproduction, **conjugation** does involve the transfer of hereditary material from one cell to another by cell-to-cell contact. As a result, there is a **recombination** (Chapter 11) of genetic material in the recipient cell. Conjugation occurs among prokaryotes and unicellular eukaryotes.

Meiotic Cell Division

Meiotic cell division occurs only in eukaryotic organisms that are either diploid or have some higher multiple of the diploid number of chromosomes (**polyploidy**) at prophase of cell division. **Diploidy** (*2n*, where *n* equals the number of different chromosomes) is the condition of having two sets of chromosomes. Chromosomes from each set having the same kind of genetic information are termed **homologous chromosomes (homologues)**. For example, the somatic cells (all the body cells except those giving rise to the gametes) of human beings possess 23 pairs of homologues, or 46 chromosomes. **Meiosis**, a form of karyokinesis, is the separation of homologues and their distribution to different developing daughter cells. The resulting daughter cells possess half the diploid number of chromosomes, a condition termed **haploidy (monoploidy;** *n*). The production of haploid gametes (sex cells) from meiotic cell division (also termed **reduction division**) leads to the formation of diploid zygotes at fertilization. If the gametes were *not* haploid, the chromosome number would double with each succeeding generation (from 46 to 92, 184, 368, . . .).

Meiotically dividing cells divide twice to produce four cells (as opposed to the single division in mitotic cell division). DNA replication occurs only once, however, prior to the first division and is the reason for the haploid condition of the four daughter cells.

Another difference between mitotic and meiotic cell divisions is that the haploid products of reduction division have new combinations of genetic information. This is a consequence of a process termed *crossing over*, during which homologous chromosomes exchange homologous portions of themselves. The result is a *recombination* of hereditary information.

Meiosis resembles mitosis in that

- it occurs concurrently with cytokinesis
- it can be described in terms of the same descriptive phases as in mitosis
- much of the mechanics (e.g., spindle apparatus formation and migration of chromosomes) is the same

Premeiotic interphase is not markedly different from that preceding mitotic cell division. Homologous chromatin filaments are replicated so that each homologue is transformed into two identical nucleoproteins attached at their centromeres.

Prophase I of meiosis is more complex than mitotic prophase. It is divided up into five stages (Figure 8.7b-e):

leptotene — (condensation) the replicated chromatin filaments undergo superwinding and superfolding to form chromosomes

zygotene — (synapsis or pairing) homologous chromosomes come together congruently along their lengths so as to form a tetrad of chromatids

pachytene — (recombination or crossing over) nonsister chromatids of homologous chromosomes break and exchange homologous segments

diplotene — (transcription) the chromatids of each tetrad separate slightly permitting gene transcription and the assembly and stockpiling of RNA for use, ultimately, in the zygote

diakinesis — (recondensation) homologous chromatids reassume the tight tetrad formation

Toward the end of prophase I, the spindle apparatus is formed, and the microtubular spindle and the nucleolus seem to disperse and disappear.

In **metaphase I**, tetrads organize at the center of the microtubular spindle, parallel to the equatorial plane (Figure 8.7f). The homologous chromosomes of each tetrad are oriented so that they are on opposite sides of the plane. (The side on which a given homologue is positioned is based on random distribution.)

During **anaphase I**, the homologous chromosomes of each tetrad separate and migrate to opposite ends of the spindle apparatus (Figure 8.7g). **Telophase I** marks

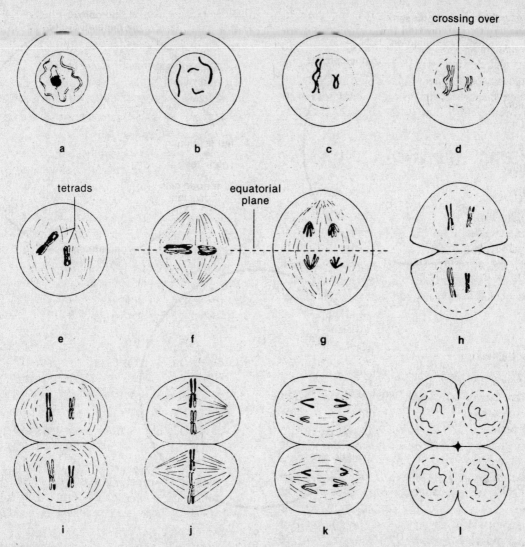

crossing over

a   b   c   d

tetrads

equatorial plane

e   f   g   h

i   j   k   l

Figure 8.7. Generalized diploid eukaryotic cell in stages of meiotic cell division: (a) G_2 of interphase, (b) leptotene, (c) zygotene, (d) pachytene, (e) diakinesis, (f) metaphase I, (g) anaphase I, (h) telophase I, (i) prophase II, (j) metaphase II, (k) anaphase II, and (l) telophase II. The diplotene stage of meiotic interphase has been deleted because of the inability to illustrate it in sufficient detail.

the end of the first meiotic division. The homologues have arrived at opposite poles of the dividing cell. *Note:* Each group of chromosomes is haploid in number but contains a diploid amount of DNA. Depending on the species, new nuclear envelopes and nucleoli may or may not form, cytokinesis may or may not take place, and interphase may or may not occur. Even if interphase does occur between the first and second meiotic divisions, no DNA replication occurs.

The primary function of the second meiotic division is the separation of sister chromatids and their segregation to daughter cells. The events of meiosis II are principally the same as those of mitosis (Figure 8.7i-l). The resultant daughter cells are now haploid, that is, each cell contains only one set of hereditary material.

Life Cycles

Meiotic cell division occurs at different times during the life cycles of different eukaryotic organisms. The variations can be categorized into three major types (Figure 8.8). In the life cycle of fungi, some algae, and a few protistans, meiotic cell division takes place immediately after fertilization. Four haploid cells are produced that eventually develop into spores. The spores mitotically divide (**germination**), developing into the **gametophyte** (haploid) **generation**, the dominant phase of the life cycle. The **gametophytes** eventually produce gametes that fuse to form a diploid zygote. The diploid cell soon undergoes meiotic cell division, thus beginning the cycle again.

In many plants, including all higher plants, meiotic

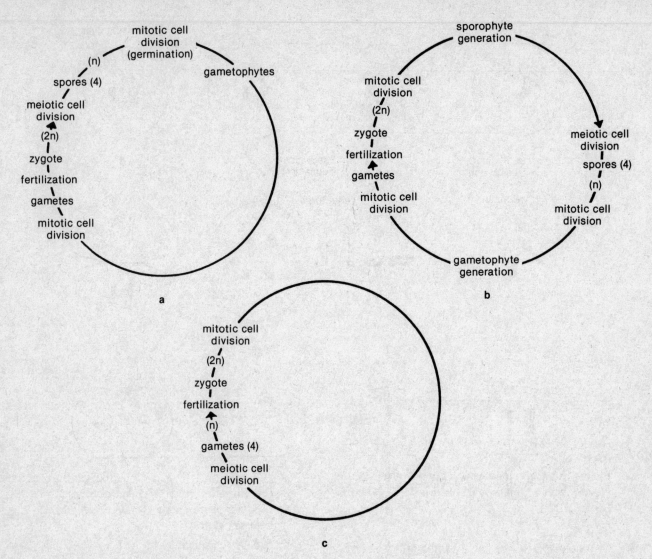

Figure 8.8. The life cycles of: (a) fungi, some algae, and a few protistans, (b) higher plants, and (c) many protistans and some lower plants.

cell division occurs at an intermediate point in their life cycle. Within their life cycles, these organisms alternate between haploid and diploid generations (alternations of generations). Fertilization initiates the **sporophyte** (diploid) **generation.** **Sporophytes** grow and develop through mitotic cell division, and at maturity, produce asexual reproductive cells, **meiospores,** through meiotic cell division (**sporogenesis**). Meiospores undergo mitotic cell division and give rise to the gametophyte (haploid) generation. Special gametophyte cells differentiate into eggs or sperm. Fusion of complementary gametes (fertilization) returns the cycle to the sporophyte generation.

Meiotic cell division in animals, many protistans, and some plants also involves gametogenesis. Gamete

formation, however, occurs only in special reproductive tissues or organs (**gonads**) in the male and female. In the males, **spermatogenesis** (Figure 8.9a) results in the formation of four haploid cells that differentiate into **sperm.** In females, **öogenesis** (Figure 8.9b) usually results in the formation of only one egg, or **ovum.** The other two or three nuclei of meiosis are discarded within polar bodies, thus maximizing the cytoplasmic content of the egg cell. Fertilization results in the formation of a diploid zygote. As a result of mitotic cell division, the zygote grows and differentiates into the diploid individual. This is the dominant phase of the life cycle. Eventually, the diploid organism produces gametes and the life cycle may be repeated once more.

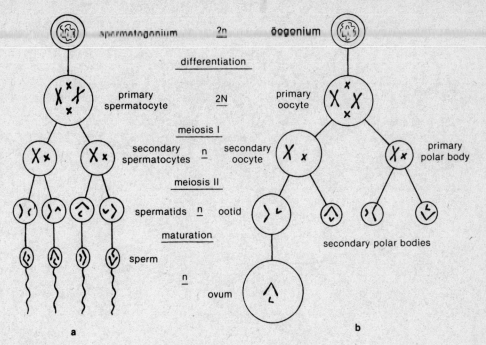

Figure 8.9. Animal gametogenesis: (a) spermatogenesis and (b) öogenesis. *Note:* In many species, the primary polar body does not divide.

CHAPTER REVIEW

1. Contrast asexual reproduction with sexual reproduction.

2. Define the following terms: clone, differentiation, gamete, gametogenesis, fertilization, and zygote.

3. Outline and describe the events of prokaryotic fission.

4. Outline and describe the events of the eukaryotic cell cycle.

5. Compare and contrast mitotic and meiotic cell division. Give strong consideration to the specific events of each.

6. Describe the structure of the metaphase chromosome of mitosis.

7. Contrast cytokinesis as it occurs in animal and plant cells.

8. Describe each of the following forms of asexual reproduction: budding, sporulation, fragmentation, regeneration, vegetative reproduction, and parthenogenesis.

9. What is conjugation?

10. Describe the three major forms of eukaryotic life cycles.

11. Compare and contrast spermatogenesis with öogenesis.

HUMAN ANATOMY AND PHYSIOLOGY

Human anatomy and physiology studies the structure and function of the human body. Cells of the human body are as varied in shape as in function (Chapter 4). Cell function is determined by the active genetic material within each cell. Cells of similar structure and function are organized into **tissues**. Several tissues performing a special function make up an **organ**. In turn, organs are organized into **organ systems**, which perform specific and complex body functions — movement, support, maintenence, and reproduction, for example.

Ideally, the human body functions properly in an ever-changing external environment. To maintain internal physiological stability, the body must adjust internal conditions in response to external conditions. This equilibrium, or homeostasis, involves the coordinated responses of *all* the body's organs and systems. These **adaptive responses ensure** survival of the individual *and* survival of the species.

TISSUES

Tissues are classified into four basic types: epithelial, muscle, connective, and nerve tissues. The size, shape, and arrangement of cells varies from tissue to tissue.

Epithelial Tissue

The cells of **epithelial tissues (epithelium)** are compactly arranged, and little or no extracellular material surrounds them. Epithelial tissues cover free surfaces inside and outside the body (Table 9.1).

Epithelial tissues are divided into three types. **Simple squamous** epithelial tissue is one cell layer thick, a flat structure that permits easy diffusion of substances. **Stratified squamous** epithelium contains several layers of cells that serve to protect underlying tissues. **Simple columnar** epithelium contains two cell types arranged in a single layer: *columnar cells* (function in absorption) and *goblet* cells (secrete mucus).

Muscle Tissue

The main function of **muscle tissue** is contraction. Location, microscopic appearance, and nervous control are the criteria that further subdivide muscle tissue (Table 9.2).

Skeletal Muscle. Skeletal muscle is composed of long, narrow threadlike cells, 0.01–0.1 mm in diameter and varying to more than 3 cm in length. This type of muscle is attached to bones and is innervated by the somatic nervous system. Regular cross-striations appear in each cell. Skeletal muscle is also known as *striated voluntary muscle*.

Cardiac Muscle. Cardiac muscle is found in the walls of the heart. It is innervated by the autonomic nervous system. Cardiac muscle cells branch and anastomose (join together) but are not one continuous mass of cytoplasm. Unique dark bands, *intercalated disks*, appear at the ends of cells. Cardiac muscle is also known as *striated involuntary muscle.*

Smooth or Visceral Muscle. No striations appear in the cells of **smooth muscle**. These cells are less than 500 microns in length. Like cardiac muscle, smooth

Table 9.1. Epithelial Tissues.

Tissue	Simple squamous	Stratified squamous	Simple columnar
Microscopic Appearance			

Location	alveoli of lungs
	lining of blood and lymphatic
	  vessels
	pericardium
	peritoneum
	pleura
		epidermis of skin
		lining of mouth and
		  esophagus
			lining of stomach and
			  intestines
			lining of respiratory tract

Function	diffusion filtration osmosis	protection	protection secretion absorption

Table 9.2. Muscle Tissues.

Tissue	Microscopic Appearance	Location	Function
Skeletal muscle		attached to bones eyeball muscles upper portion of esophagus	move bones and eyes swallowing

Visceral muscle		iris
		ducts of glands
		walls of: respiratory,
		  digestive, genito-urinary
		  systems
		blood and lymphatic vessels
			change diameter of pupil
			move substances along
			change diameter (regulation
			  of blood pressure)

Cardiac muscle		walls of heart	contraction

muscle is controlled by the autonomic nervous system. It is found in blood vessels, the intestines, the uterus, and in the walls of other hollow internal structures. Smooth muscle is known as *nonstriated involuntary muscle*.

Connective Tissue

Connective tissue is widespread throughout the body. Extracellular materials make up most of the connective tissue and determine the types of connective tissue. The cells of connective tissue appear to function in the formation and maintenance of the extracellular material (Table 9.3).

Nerve Tissue

Table 9.4 summarizes structure, location and function of nerve tissues. **Neurons**, the cells that conduct nerve impulses, are divided into *sensory (afferent)* neurons and *motor (efferent)* neurons. The former transmit an impulse from the body to the spinal cord or brain. The latter carry an impulse away from the brain and spinal cord. **Interneurons** function as connectors between the two. Little is known about normal functions of the neuroglia.

ORGANS AND ORGAN SYSTEMS

Organs are composed of different tissues combined into a structure that contributes to the functioning and survival of the individual. Single organs, however, do not act alone to contribute to bodily functions; they work in conjunction with other organs to form systems. Integration of organ systems assures efficient controlled responses to the body's environment. These responses are: support and movement, control and integration, body maintenance, and reproduction.

SUPPORT AND MOVEMENT

The skeletal and muscular systems are the primary providers of the body's movement and support functions.

Skeletal System

The **skeletal system** of bones and the joints connecting those bones serves as a series of levers and fulcrums within the body (Figure 9.1). Bone, cartilage, and hemopoietic connective tissues are the tissue types found within this system.

Bones are classified according to their shape. *Long bones* include the humerus, radius, ulna (arm bones), phalanges (foot and hand bones), and the femur, tibia, and fibula (leg bones). The wrist (carpals) and ankle (tarsals) bones are *short bones*. The frontal and parietal bones of the skull, ribs, and scapulae (shoulder bones) are *flat bones*. *Irregular-shaped bones* include vertebrae, the patella (knee cap), the sacrum and coccyx (tailbones), and the mandible and maxilla·(jawbones), among others.

Bone is more than a hard, lifeless structure. **Osteocytes**, bone cells, are embedded in a collagen and

Table 9.3. Connective Tissues.

Tissue Type	Description	Location	Function
reticular	3-dimensional web	spleen, lymph nodes bone marrow	filtering, phagocytosis
aerolar	loose, ordinary	between tissues, around organs	connection
adipose	fat	under skin, around organs	protection, insulation, food reserves
dense fibrous	collagenous, elastic	tendons, ligaments, dermis of skin, scars	strong flexible support
bone	calcified	skeleton	protection support
cartilage	collagenous and gelled	nasal septum, external ear, eustachian tube, larnyx, tracheal and bronchial rings, vertebral discs	firm flexible support
hemopoietic	blood cell forming	lymphatic system vessels	protection, formation of blood cells
blood		in blood	protection, transportation

Table 9.4. Nerve Tissue.

Types	Neurons	Neuroglia
Microscopic Appearance		
Location	nerves brain spinal cord	brain spinal cord
Function	impulse conduction	support — anchored to blood vessels protection — develop myelin sheath

calcium matrix that gives bone both flexibility and strength. Within the periosteum (the dense fibrous membrane covering the bone) of growing bones are the **osteoblasts,** bone-forming cells. The periosteum also serves as a place of attachment for tendons and ligaments.

The embryonic skeleton consists of hyaline cartilage (shaped like bones) that contains osteoblasts. Mucopolysaccharides, carbohydrates secreted by the Golgi apparatus, and collagen, protein secreted by the endoplasmic reticulum, accumulate around the osteoblasts. As calcium salts from blood are deposited into this bone matrix, the cartilage hardens, forming bones. **Ossification (osteogenesis)** is the process of bone hardening.

Bone growth results from the resorption of the bone cavity by *osteoclasts,* followed by laying down and calcification of new bone on the outside by action of osteoblasts. During childhood, bone formation proceeds at a rate faster than bone destruction, resulting in skeletal growth. Rates of growth and destruction balance during middle adult years. During old age, the rate of bone resorption is greater than formation, causing actual shrinkage in some cases. This smaller bone becomes more porous and fragile resulting in *senile osteoporosis.*

The human skeleton is divided into two parts. The **axial skeleton** — the upright part of the body — consists of the vertebral column, sternum, ribs, the skull, and hyoid process. The **appendicular skeleton** is composed of those bones attached as appendages.

The **vertebral column** (spine) is the segmented flexible axis of the body. It encloses the spinal cord. The vertebrae are classed according to their position. The upper seven *cervical vertebrae* are found in the neck region. Twelve *thoracic vertebrae* are located behind the thorax region of the body, followed by five *lumbar vertebrae* of the lower back. The *sacrum* and *coccyx* are formed by the fusion of vertebrae and are attached below the lumbar vertebrae. Together with the thoracic vertebrae, the bones of the sternum and ribs form the **thorax,** the bony cage protecting the heart and lungs. The upper seven pairs of *true ribs* fasten to the sternum via costal cartilage. The remaining five pairs of *false ribs* do not attach directly to the sternum. The upper three pairs attach through the costal cartilage of the seventh true rib. The last two pairs are *floating ribs,* unattached.

The 28 bones of the skull function primarily as protective structures. Except for the *mandible,* the skull bones are immovable. They are joined by fibrous joints. The two major portions of the skull are the cranium (brain case) and the face. The *hyoid process* is a U-shaped single bone in the neck. This is the only bone in the human body that is not joined (articulated) with any other bone.

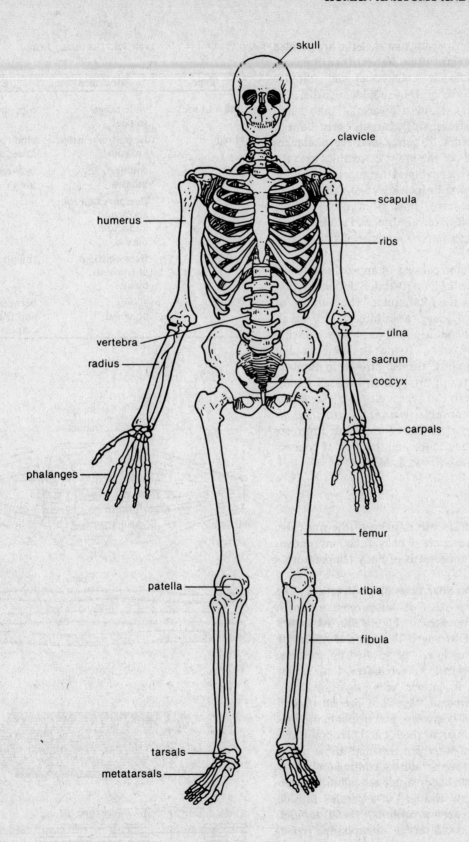

Figure 9.1. The human skeleton.

The bones of the appendicular skeleton are divided into upper and lower extremities. Bones of the upper extremeties are: the clavicle (collarbone) and scapulae (shoulder bones) forming the shoulder girdle; the humerus (upper arm), radius (forearm, thumb side), ulna (forearm, little finger side), carpals (wrist bones), metacarpals (framework of palm), and the phalanges (finger bones). Bones of the lower extremities are: the innominate bones (large hip bones that together with the sacrum and coccyx form the pelvic girdle), femur (thigh bone), patella (knee cap), tibia (shin bone), fibula (lateral side of lower leg), tarsals (heel and ankle bones), metatarsals (long bones in feet), and the phalanges (toe bones).

The human skeleton consists of approximately 206 bones. The presence of two additional bone types, however, may affect the total count. The number of small round *sesamoid bones* varies in each individual. These bones are found in tendons where pressure develops. *Wormian bones*, found in some cranial sutures (immovable joints), are also variable in number.

A joint is the area of articulation between bones. They are classified into two types: movable *diarthrotic articulations*, and immovable *synarthrotic articulations* without a joint cavity. Diarthroses are divided into six classes based upon their shape and the type of connective tissue in each, as shown in Table 9.5.

Muscular System

The skeletal and smooth muscles of the muscular system serve three generalized body functions: movement, posture, and homeostasis of body temperature.

Cellular and Molecular Basis of Contraction. The cellular and molecular basis of muscle contraction in skeletal muscle is presented in Figure 9.2. **Myofibril** contractile fibers are arranged in a regular repeating pattern giving the muscle cell its striated appearance. Each repeating unit is called a **sarcomere**. Four major proteins are found in muscle cells, *myosin, actin, tropomyosin,* and *troponin*. Myosin is present only in the *A band*. Actin, tropomyosin, and troponin are present in the *I band* and part of the A band. Troponin and tropomyosin are regulatory proteins that allow the A and I bands to slide together during contraction.

A highly modified endoplasmic reticulum, the **sarcoplasmic reticulum**, is arranged in a parallel fashion filling the spaces between myofibrils. The *T tubules*, sandwiched between two layers of sarcoplasmic reticulum, are invaginations of the *sarcolemma* sheath surrounding the myofibrils. This constitutes the *Z lines*. A nerve impulse from the neuromuscular junction travels throughout the sarcolemma via the T tubule system. A

Table 9.5. Diarthrotic Joints.

Type	Movement	Examples
Ball and socket	wide range triaxial	hip, shoulders
Hinge	flexion/extension, uniaxial	elbow, fingers, knee, ankle
Pivot	rotation, uniaxial	between radius and ulna
Ellipsoid	flexion/extension, abduction/ adduction, biaxial	wrist
Saddle	freer ellipsoid movements, biaxial	thumb
Gliding	gliding, nonaxial	between sacrum and iliac, between carpals

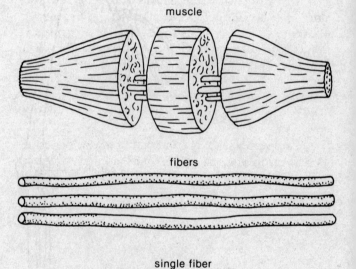

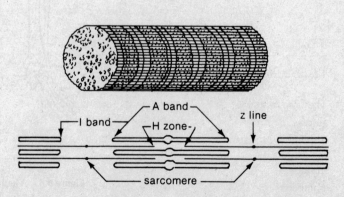

Figure 9.2. Cellular and molecular structure of muscle.

release of calcium ions from the sarcoplasmic reticulum is initiated which, in turn, initiates muscle contraction. Calcium binds to the troponin portion of the thin filament in the I band, changing the shape of troponin and tropomyosin. Actin and myosin are then free to form cross bridges. Hydrolysis of myosin-bound ATP generates the energy needed for the actin molecule to move along the myosin molecule. The sliding action shortens the sarcomere; the myofibril shortens; the muscle fibers shorten. Muscle contraction occurs when a significant number of these fibers contract.

Relaxation occurs through a reversal of these steps. Release of calcium from the troponin molecule inhibits cross bridge formation. The calcium is sequestered into the sarcoplasmic reticulum.

Skeletal Muscle. Contraction of skeletal muscle requires a nerve impulse. The type of contraction depends on the number of nerve fibers innervating the muscle, the number of nerve impulses per unit of time, the time elapsed before response occurs, and the coordination of fiber contractions. In response to stimulation, movement occurs as muscles pull on bones.

The molecular arrangement within the myofibril does not permit muscle lengthening. Muscles must therefore work in groups to accomplish movement — contraction of one muscle occurs concomitantly with the relaxation of another. Reversal of this process returns the muscles to their original state. Skeletal muscles are grouped according to their actions (Table 9.6).

Cardiac Muscle. The internal structure and function of cardiac muscle cells is similar to that of skeletal muscle. Cardiac muscle is discussed later in this chapter under "Cardiovascular System."

Table 9.6. Skeletal Muscle Groups and Their Actions.

Muscle Group	Action
flexor/extensor	change angle of joint (decrease/increase)
abductor/adductor	move away/toward midline
levator/depressor	raise/lower
supinator/pronator	turn palm of hand upward/downward
rotator	pivot on axis
sphincter	decrease size of opening
tensor	make rigid

Smooth Muscle. Smooth muscle differs radically from skeletal muscle. Actin, but not myosin is present. Cells are surrounded by a sarcolemma and glycoprotein. Those junctions lacking this glycoprotein coat serve as transmission points for the relay of nerve action potentials. The sarcoplasmic reticulum is poorly developed.

Unitary muscles occur in sheets or layers, and contract spontaneously. This type of smooth muscle is very responsive to mechanical stretch stimuli, and is found in the gastrointestinal tract and in ureters. Multi-unit muscles, in contrast, are innervated by motor nerves. These muscles do not contract spontaneously and do not respond to stretch stimuli. They may be arranged in sheets, bundles, or as single cells. Multi-unit muscle is typically found in blood vessels, in the spleen, and in the muscles that attach to hair and cause the "goose bump" reflex.

Smooth muscle functions by slow, rhythmic contractions. This muscular activity is responsible for maintenance of bodily functions — respiration, circulation of blood, digestion, absorption, and excretion.

CONTROL AND INTEGRATION

The nervous and endocrine systems relay messages within the body. The nervous system sends electrical impulses, whereas the endocrine system sends chemical messages (hormones). These messages generate not only single responses, but function in the control and integration of the entire body.

Nervous System

Neuron Structure. Neurons, the impulse conducting cells of the nervous system are composed of three parts: the **soma** or cell body, the **axon,** a single extension from the soma, and **dendrites,** the branching portions (Table 9.4). The soma resembles other body cells. Axons vary from less than 1 mm to more than a meter in length. The diameter of the neuron is determined by the amount of *myelin sheath* (protein and lipid) laid down in concentric rings by *Schwann cells.* The axon carries the nerve impulse away from the cell body. Dendrites receive a nerve impulse across the **synapse** (junction point of two neurons).

Conduction of Nerve Impulses. A nerve impulse is the propagation of electrical disturbance along the length of a nerve cell. Impulses are generated by stimulation that causes a change in the plasma membrane's sodium permeability. Depolarization of the membrane occurs as sodium pours into the cell and the internal concentration of positive charges is increased.

The deficit of positive charges on the outside results in a surface negativity. A local current develops between the negatively charged portion of the membrane and the positive portion directly adjacent to it. This adjacent membrane becomes depolarized and the original site reverses its potential and is repolarized. This **action potential** is carried along the plasma membrane of the neuron. The thickness of the neuron's myelin sheath determines the speed of nerve impulse conduction. Impulses may travel at a rate up to 100 meters per second.

Conduction of the nerve impulse from cell to cell involves a different mechanism. The terminals, or *synaptic knobs* at the end of each axon contain **neurotransmitter** molecules stored in vesicles. As the action potential reaches the synaptic space between two cells, neurotransmitter molecules are released. Diffusion across the plasma membrane to the adjacent cell produces a response. The form of response is based on the receiving cell type: neuron cells depolarize, muscle cells contract, and glands secrete. **Acetylcholine** is the neurotransmitter that operates primarily on skeletal muscle. *Norepinephrine* acts at glandular, cardiac, and smooth muscle junctions.

Reflex Arcs. The human body constantly adapts to a changing environment. A stimulus begins at a **receptor** (sensory nerve ending) and results in changes of body **effectors** (muscles and glands). The travel of a nerve impulse from a sensory receptor to the spinal cord or brain and on to an effector organ is known as a **reflex arc**. Many, but not all, nerve impulses are conducted via this route.

In a **short reflex arc**, the sensory (afferent) neuron enters the dorsal root of the spinal cord and leaves through the ventral root of the same segment (Figure 9.3). However, the majority of afferent fibers split into ascending and descending branches soon after entering the spinal cord. These branches travel varying lengths before connecting with *internuncial neurons* that pass the impulse to a number of motor (efferent) neurons. This path, known as a **long reflex arc**, allows a single afferent neuron to activate many efferent neurons. Likewise, a single efferent neuron can receive impulses from several afferent sources. This vast interconnecting system of neurons permits integration of messages and coordination of the body's response.

Somatic Nervous System. Communication between skeletal muscle and other parts of the body is carried through the **somatic nervous system**, via the brain, spinal cord and cranial and spinal nerves.

The brain is divided into: the **cerebrum**, the **diencephalon**, the **cerebellum**, the **medulla oblongata**, the **pons**, and the **mesencephalon** or **midbrain**. The

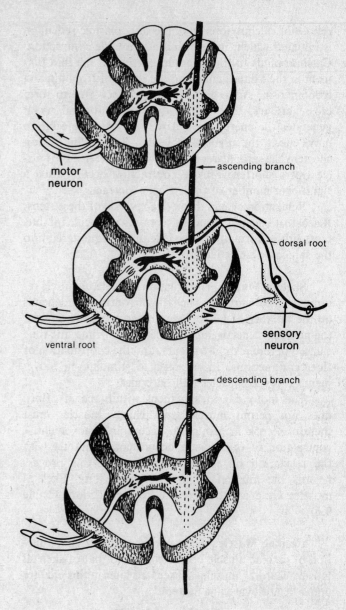

Figure 9.3. Reflex arc.

medulla, pons, and midbrain are commonly called the *brainstem*. See Table 9.7 for a summary of the structures and their functions.

The spinal cord conducts impulses back and forth between the brain and peripheral nerves. It also serves as a reflex center for all reflexes except those of the face. An action that results from the passage of a nerve impulse along a reflex arc is known as a reflex — an involuntary response that does not involve activity of the cerebral cortex.

The brain and spinal cord are protected by the bony cranium and vertebrae, the membranous **meninges**, and the **cerebrospinal fluid** that cushions against jarring. The meninges have three layers: the outer fibrous *dura mater*, the innermost transparent *pia mater*, and the

Table 9.7. Brain Functions.

Structure	Functions	Structure	Functions
Cerebrum	consciousness memory speech emotion movements spatial awareness sight hearing general sensations	Cerebellum	maintain equilibrium control muscle action
		Medulla oblongata	control respiration, blood pressure and heartbeat mediate vomiting and coughing reflexes pass impulses between brain and spinal cord
Diencephalon (thalamus and hyothalamus)	relay impulses to cerebrum produce complex reflexes synthesize hormones regulate body temperature regulate appetite	Pons	houses cranial nerves passes nerve fibers between brain and spinal cord
		Mesencephalon or Midbrain	center for cranial nerves; control eye movement and pupil reflexes

delicate *arachnoid membrane* between them. Blood vessels are contained within the pia mater. Cerebrospinal fluid is produced by a network of capillaries (the choroid plexus) and fills the ventricles of the brain, the subarachnoid space, and the central spinal cord canal. It also separates the arachnoid membrane and the pia mater. This protective cushioning fluid circulates through the ventricles and is absorbed back into the venous blood supply.

Autonomic Nervous System. The **autonomic nervous system** supplies innervation to those responses termed "involuntary." Consisting only of efferent neurons, it supplies body-maintaining organs, many thoracic and abdominal organs, glands, hair muscles, and the iris and ciliary muscles. Responses of the autonomic nervous system can be triggered by afferent neurons through long reflex arcs.

Sympathetic ganglia lie anteriorly on either side of the spinal column. Short fibers connect them and give the appearance of beads on a chain. They are referred to as "sympathetic chain ganglia." In contrast, **parasympathetic ganglia** lie near or within their effector organ.

Preganglionic neurons carry impulses from the spinal cord to the ganglia. Synapses within the ganglia pass the impulse to **postganglionic neurons**, which conduct the impulse to the effector. Preganglionic axons of the sympathetic system synapse with *many* postganglionic neurons, which in turn, travel to *many* organs. This explains the widespread physiologic responses triggered by the sympathetic system. Postganglionic sympathetic axons often form a complicated plexus (network) before terminating at an effector organ. Preganglionic para-

sympathetic neurons, however, synapse with postganglionic neurons to single effector organs.

The majority of effector organs are innervated by both sympathetic and parasympathetic neurons. Exceptions are the sweat glands, hair muscles, and most blood vessels, which receive only sympathetic input. Organs under dual innervation receive continual impulses. Acetylcholine and norepinephrine, the chemical transmittors, act antagonistically; that is, one chemical may cause the organ to increase activity, whereas the other may cause the organ to decrease activity. The sum of these two opposing responses determines the overall response. (Learning occurs within both somatic and autonomic nervous systems.)

Endocrine System

Chemical messages, hormones produced by the endocrine glands are carried through the blood. Table 9.8 lists the endocrine glands, the hormones they produce, and hormone functions.

Hormones are protein molecules or steroid molecules. The importance of hormones cannot be exaggerated. They affect nearly every cell within the body. Hormones affect target cells — cells that have hormone-specific receptors on their cell surface. The hormone-receptor complex is actively transported into the cell. The complex dissociates, and the hormone moves into the nucleus and binds to hormone-specific proteins on the DNA. This causes the cell to respond in a specialized manner by triggering DNA replication or mRNA transcription.

Table 9.8. The Endocrine Glands.

Gland	Location	Hormone Produced	Function
Pituitary gland *Anterior lobe*	Cranium		
		Somatotropin (STH)	promotes body growth
		Prolactin	milk secretion
		Thyroid stimulating hormone (TSH)	maintenance of and secretion from thyroid
		Adrenocorticotropin (ACTH)	maintenance of and secretion from adrenal cortex
		Follicle stimulating hormone (FSH)	*females* — maturation of follicles (egg-producing cells) *males* — maintenance of seminiferous tubules and spermatogenesis
		Luteinizing hormone (LH)	ovulation, stimulates secretion by corpus luteum
		Interstitial cell-stimulating hormone (ICSH)	stimulates production of testosterone in testicular Leydig cells
		Melanocyte-stimulating hormone (MSH)	increases pigmentation of the skin
Posterior lobe		Antidiuretic hormone (ADH)	prevents excessive urine loss
		Oxytocin	uterine contractions during labor
Pineal gland	Cranium	Melatonin	inhibits LH secretion
Thyroid	Neck	Thyroxine and triiodothyronine	regulation of metabolic rate, growth, and tissue differentiation
		Thyrocalcitonin	decreases blood calcium concentration
Parathyroid gland	Neck	Parathyroid hormone	homeostasis of blood calcium
Adrenal gland *Adrenal cortex*	Atop kidneys	Glucocorticoids	increases breakdown of proteins; increases use of fats for energy production; maintains blood pressure; anti-inflammatory effect; decreases number of blood lymphocytes and plasma cells
		Mineralocorticoids	regulates electrolytes (Na^+, K^+)
		Sex hormones	supports sexual behavior
Adrenal medulla		Epinephrine and norepinephrine	similar to sympathetic stimulation on glands, smooth muscle, and cardiac muscle
Isles of Langerhans	Pancreas	Insulin	promotes metabolism of glucose, amino acids, and fatty acids
		Glucagon	increases blood glucose concentration
Ovaries	Pelvic cavity	Estrogen and progesterone	reproduction and secondary sex characteristics
Testes	Scrotum	Testosterone	reproduction and secondary sex characteristics
Placenta (temporary endocrine gland)	Pregnant uterus	Chorionic gonadotropin, estrogen, and progesterone	maintenance of pregnancy

BODY MAINTENANCE

Maintenance, metabolism, and "housekeeping" are carried out by the respiratory, cardiovascular, digestive, and excretory systems.

Respiratory System

Even though the body has reserves of metabolic energy-supplying materials, a continual oxygen supply must be brought in from the environment. This is the function of the **respiratory system** (Figure 9.4). The

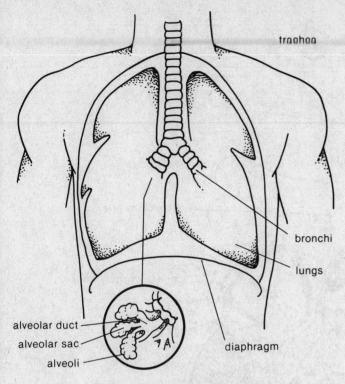

alveolar duct
alveolar sac
alveoli

trachea

bronchi

lungs

diaphragm

Figure 9.4. The respiratory system.

nose, pharynx (throat), larynx (voice box), trachea, bronchi, and lungs function in air distribution. Alveoli within the lungs function as gas exchangers, removing carbon dioxide waste from the blood and replenishing it with oxygen.

The venous blood within the capillaries of the lung pass over the microscopic alveoli. Oxygen does not directly diffuse into the plasma (solubility of oxygen in plasma is too low to meet the body's needs). Instead, four oxygen molecules reversibly bind to hemoglobin, an iron-containing protein molecule found within red blood cells. This oxygen-rich blood is carried back to the heart and is pumped throughout the body. As the capillary blood supply reaches tissues, the oxygen is released from the hemoglobin molecule. Then the oxygen diffuses out of the blood cells into the plasma and into the cytoplasm of the tissue cells.

Cardiovascular System

The **cardiovascular system** serves as the body's transportation system, transporting energy sources, oxygen, waste products, and hormones.

Blood. Blood is composed of plasma (the fluid portion) and three types of blood cells: erythrocytes (red blood cells), leukocytes (white blood cells), and thrombocytes (platelets). Plasma consists of 90 percent water, 6 to 8 percent proteins, and additional dissolved sub-

stances (electrolytes, foods, waste materials, hormones, and respiratory gases).

Erythrocytes function as oxygen carriers. Mature red blood cells are *enucleate* (without a nucleus). New cells are produced from stem cells within the bone marrow. The average life span of a circulating red blood cell is 120 days. Thereafter, they are phagocytosed (engulfed and destroyed) by reticuloendothelial cells of the liver.

Leukocytes function in the body's defense mechanisms. These mobile cells are able to squeeze out of capillaries into tissue spaces where they phagocytose microorganisms and cellular debris resulting from injury. **Lymphocytes**, highly specialized white blood cells, function in the formation of antibodies. Leukocytes originate in bone marrow and lymphatic tissue.

Thrombocytes function in blood clotting. They are formed in the marrow of bones.

The heart. A loose fitting envelope of connective tissue, the **pericardium**, covers the heart and protects it from friction. The heart itself is divided into four chambers — two upper atria and two lower ventricles (Figure 9.5). The body's venous blood supply enters the **vena cava** and flows into the **right atrium**. From here it is pumped into the **right ventricle**. Valves separating the atrial and ventricular cavities prevent blood from flowing backwards. From the right ventricle the blood is pumped via the **pulmonary artery** to the lungs. (The pulmonary artery is the only in the body that carries unoxygenated blood.) Gas exchange takes place in the lungs, and the resulting oxygenated blood is sent to the **left atrium** of the heart via **pulmonary veins**. From the left atrium, it is pumped to the **left ventricle** where the chamber walls are thickest. These walls contract and force the blood through the aorta and on to the rest of the body.

Atrial and ventricular pumping is coordinated by the cardiac pacemakers. The **sino-atrial (S-A) node**, located in the right atrium, generates electrical impulses that spread through the atria causing their contraction. This impulse triggers the **atrio-ventricular (A-V) node**. A second electrical impulse spreads throughout the ventricles generating their contraction.

The Circulatory System. The **arteries** carrying blood away from the heart have strong elastic walls of smooth muscle and connective tissue. When a ventricular contraction sends a new surge of blood at high pressure, the blood vessels expand then return to normal diameter. This action helps force the blood along the system. The diameter of the blood vessels gradually decreases along the length of the circulatory system. Arteries lead to **arterioles**, which lead to **capillaries** — blood vessels with a diameter so narrow that

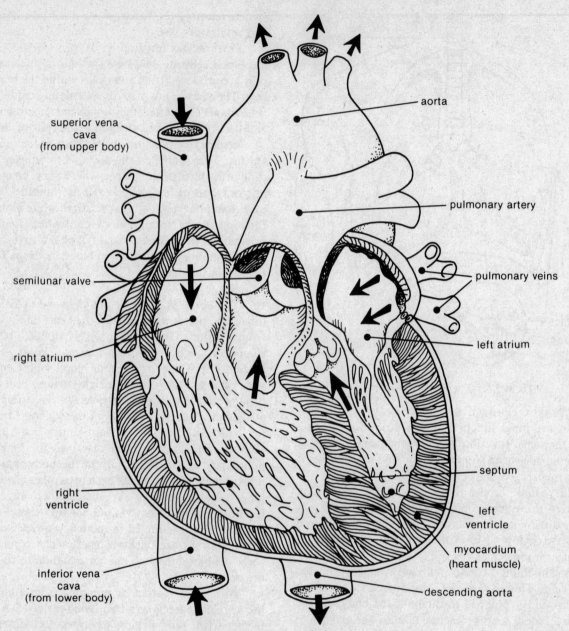

superior vena
cava
(from upper body)

aorta

pulmonary artery

semilunar valve

pulmonary veins

right atrium

left atrium

right
ventricle

septum

left
ventricle

myocardium
(heart muscle)

inferior vena
cava
(from lower body)

descending aorta

Figure 9.5. Anatomy of the heart.

single red blood cells must squeeze through. As the capillaries pass over individual tissue cells, an exchange of materials takes place. Oxygen and foodstuffs are picked up by the cell, and waste products are deposited into the blood stream. This deoxygenated or **venous blood** is pushed back to the heart through vessels of increasing size; **venules** empty into **veins** and eventually into the vena cava.

The Lymphatic System. The space between the capillaries and tissue cells is filled with interstitial fluid, or **lymph**. This fluid, containing the same materials as plasma, has filtered out of the capillaries. Some of the

fluid is resorbed into the capillaries. Additional fluid is returned to the circulatory system by way of the **lymphatic system**. Lymphatic capillaries are single layers of endothelial cells; the larger lymphatic vessels contain smooth muscle and connective tissue layers. Lymph is eventually dumped into the subclavian veins in the neck area. **Lymph nodes** are positioned along the course of the lymphatic system and function to form lymphocytes, to filter, and to dispose of large particles (including bacteria). Lymph flow is very slow compared to blood flow because no special organs push the lymph along. External forces within the body such as increase in blood flow, thoracic aspiration, and massage increase lymph flow.

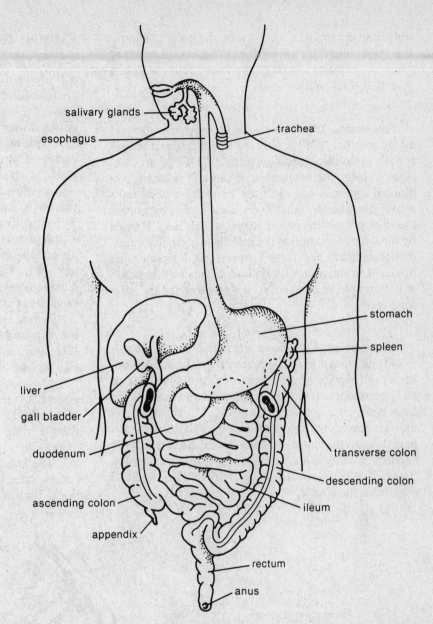

Figure 9.6. The digestive system.

Digestive System

The digestive system prepares food for absorption and use by the cells. This process involves physical and chemical alteration of foodstuffs.

Organs of digestion. The **gastrointestinal tract**, shown in Figure 9.6, consists of a tube open at both ends and accessory organs. The *mouth* (and accessory organs), *teeth*, and *salivary glands*, prepare the food for digestion by breaking it down into smaller pieces, thereby exposing more surface area to enzymatic action. *Amylase* secretions begin the chemical digestion of starches in the mouth as the *bolus* (food) is mixed by the *tongue* and pushed into the *pharynx*, the funnel-shaped connection between mouth and *esophagus*. **Peristaltic**

contractions within this tube transport the food to the *stomach* the large muscular sac that stores, grinds, and mixes the food. Glands of the stomach's inner mucosal membrane secrete mucus and *gastric juice*, which continue the chemical breakdown of the food. The exit valve, the *pyloric sphincter*, controls the rate of release of the semiliquid *chyme* into the *small intestine*. Intestinal, pancreatic, and bile secretions combine with the food in the *duodenum* (portion of the small intestine). Complex molecules of fats, proteins, carbohydrates, and nucleic acids are broken down into monomeric units and absorbed in this region. The *jejunum* and *ileum* regions secrete mucus and intestinal juice. They also absorb fat-soluble vitamins, some fatty acids and amino acids, water, bile salts, and electrolytes. The *colon* (large intestine) continues to absorb water and electrolytes. The *rectum*, the lower portion of the large intestine, stores the undigested wastes, or *feces*. A sphincter muscle of the *anus* opens the gastrointestinal tract to the outside.

Enzymatic Digestion. Most digestive enzymes are highly specific, splitting only specific chemical bonds of specific molecules. Lipase, pepsin, and HCl (hydrochloric acid) begin the breakdown of fats and proteins. Additional enzymes complete the breakdown into units readily absorbed by cells. *Trypsin* and *chymotrypsin* are the chief protein-degrading enzymes. *Bile salts*, formed by the liver and stored in the gall bladder emulsify fats, preparing them for further hydrolysis by pancreatic lipases. Carbohydrate breakdown begins in the mouth and is completed by specific enzymes secreted by *intestinal epithelium*.

Absorption. Like enzymatic digestion, the majority of food absorption takes place within the duodenum. The entire surface of the small intestine is covered with *villi*, microscopic fingerlike projections that increase the absorptive surface area (Figure 9.7). Monomeric food units, monosaccharides, amino acids, and nucleic acids pass through the single-cell epithelial layers and into the capillaries. Fatty acids, however, are transported by the lymph. The colon absorbs water, mineral salts, and bile salts from the chyme.

Excretory System

Undigested nutrients (notably cellulose), sloughed off intestinal cells, secretions, and residues of bile salts are fermented by bacteria within the large intestine. The remaining material, feces, is stored in the rectum until it passes through the anus as solid waste.

Gaseous wastes (carbon dioxide) produced by cells during respiration are removed via the lungs. Hemoglobin carries approximately ten percent of the waste carbon dioxide back to the alveoli. An additional five percent is dissolved and carried directly in the plasma. The majority of the carbon dioxide combines with water in the plasma, forming carbonic acid (H_2CO_3), which dissociates into bicarbonate ($HCO_3{}^-$) and hydrogen ions (H^+). These reactions reverse at the alveoli; carbon dioxide is released from the blood and expelled in the exhaled breath.

Liquid wastes are removed via the **sweat glands** of the skin and the urinary system. Sweat carries water, salts, and nitrogenous wastes to the body surface while also serving as a temperature regulator.

Most liquid waste is removed through the **urinary system**. Urine is formed in the *kidneys* and carried via the *ureters* to the *bladder* for storage. The *urethra* carries the urine out of the body. (The urethra is also part of the genital system in males.)

The kidneys function to filter blood, and to form, concentrate, and collect urine. The **nephron** is the functional unit of the kidney. Capillaries intermingle with

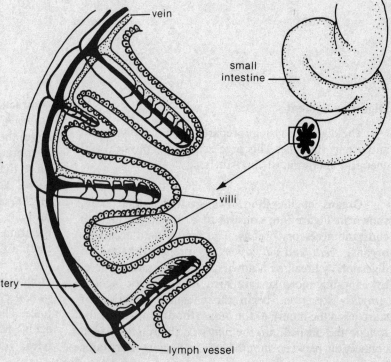

Figure 9.7. Cross section of the small intestine and villi.

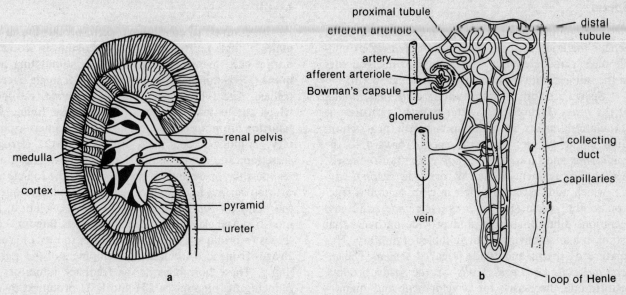

Figure 9.8. Cross section of kidney (a) and nephron (b).

kidney tubules, as shown in Figure 9.8. Plasma and dissolved substances are forced out of the capillaries. This **filtrate** enters the *Bowman's capsule*. Water and selected solutes are resorbed into the surrounding capillaries as the filtrate passes through the *proximal tubule*. Excess electrolytes and waste products remain within the nephron. Sodium is actively transported out as the filtrate passes up the *Loop of Henle*. Water is lost via osmosis. More water is lost in the *distal tubule* as the acid-base balance is adjusted and more ions are pumped in or out. The dilute urine passes into the *collecting ducts*. These ducts run parallel to the Loop of Henle. The osmotic gradient formed in that region pulls addi-

tional water out of the urine within the collecting ducts. **Antidiuretic hormone (ADH)** controls the water permeability of the collecting tubules.

REPRODUCTION

Survival of the species depends on survival of the individual *and* reproduction of those surviving individuals. Specialized reproductive cells are shown in Figure 9.9. Male reproductive cells, **sperm**, are approximately 0.05 mm in length; the female reproductive cell, the **egg** or **ovum**, is approximately 0.25 mm in diameter.

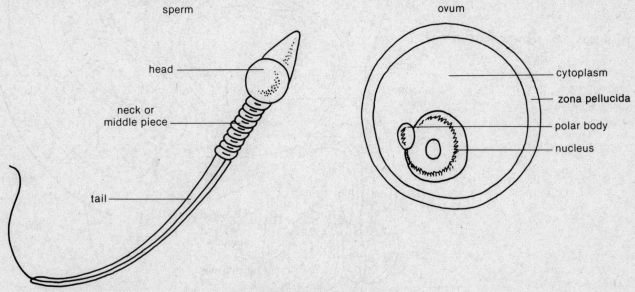

Figure 9.9. Sperm and egg cells.

Sperm

The compact head of the sperm contains the DNA genetic material. Mitochondria within the neck or middle piece provide energy. Contraction of microtubules in the tail generate movement of individual sperm.

Sperm are formed within the *seminiferous tubules* of the *testes* (Figure 9.10). After puberty, testosterone stimulation results in continuous sperm production. Cilia in the tubules move the sperm forward to the *epididymis* and on to the *vas deferens* out of the testis. A slight enlargement of the vas forms the *seminal vesicle*, which adds secretions high in citric acid and fructose to the sperm. In the *prostate gland*, additional secretions dilute the sperm and provide nutrients that establish a microenvironment suitable for motility. This mixture of sperm and fluids is called **semen**. The *interstitial cells* or *Leydig cells* of the testis produce testosterone, necessary for development and maintenance of sexual organs and secondary sexual characteristics.

Ovum

In contrast to continuous sperm production in the male, a single reproductive cell, the ovum, is produced during each menstrual cycle. **Follicle stimulating hormone (FSH)** initiates the maturation of a single ovarian follicle. The follicle produces the hormone, **estrogen,** which stimulates the growth of the uterine lining. **Luteinizing hormone (LH),** produced by the anterior pituitary, stimulates continued follicle growth, estrogen secretion, and ovum maturation. This hormone is also responsible for **ovulation**, the rupture of the follicle and ovarian surface and the escape of the ovum into one of the *fallopian tubes* (Figure 9.11). After ovulation, the modified follicle, now called the **corpus luteum**, continues to produce estrogen in addition to progesterone. Progesterone also stimulates development of the uterine lining. These hormones act as feedback inhibitors by reducing the amount of LH and FSH produced by the anterior pituitary. Reduction of LH and FSH, in turn, causes degeneration of the corpus luteum and reduction

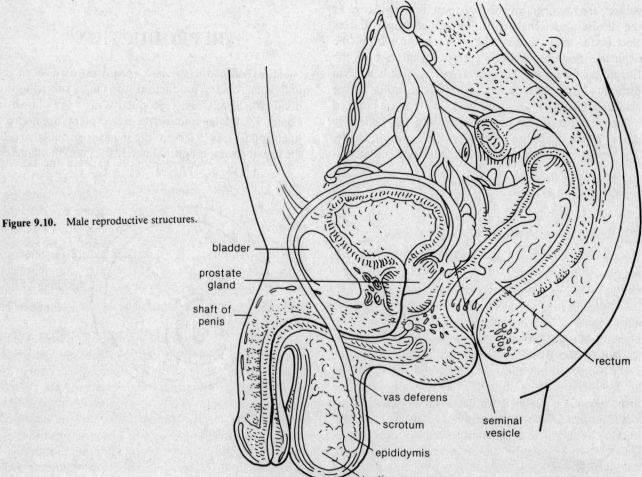

Figure 9.10. Male reproductive structures.

bladder

prostate gland

shaft of penis

rectum

vas deferens

scrotum

seminal vesicle

epididymis

testis

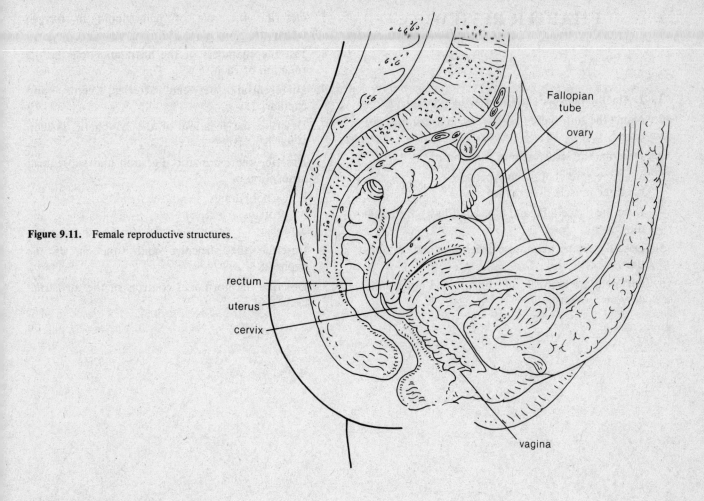

Figure 9.11. Female reproductive structures.

in production of estrogen and progesterone. This decrease in estrogen and progesterone levels causes the sloughing off of the uterine lining and the beginning of another menstrual cycle.

Fertilization and Pregnancy

During copulation, sperm are deposited near the opening of the uterus, the **cervix**. Many sperm die within the acidic vaginal environment. Significant numbers swim against opposing currents set up by uterine and oviduct cells, to eventually reach the fallopian tubes. If an egg is present, a single sperm penetrates the egg and **fertilization** is accomplished. Fertilization normally occurs within the fallopian tubes. The genetic material from each parent joins in the nucleus of the fertilized egg and a **zygote** is formed. The dividing ball of cells is termed a **blastocyst**. It travels down the fallopian tube to the uterus. Normal **implantation** occurs as the blastocyst sinks into the uterine mucosa after proteolytic (protein digesting) enzyme digestion. Implantation inhibits corpus luteum degeneration. Estrogen and pro-

gesterone levels therefore continue to increase during pregnancy. Two membranes develop from the blastocyst, the amnion and the chorion. The **amnion** forms a fluid-filled cavity in which the developing embryo lives. The **chorion** fuses with maternal tissues to form the placenta. The placenta functions to produce estrogen and progesterone as well as to serve as the site for the maternal-fetal exchange of gases, food and waste material. The *umbilical cord* houses the umbilical arteries and umbilical vein. These vessels terminate in capillaries within the placenta in a pool of maternal blood.

Full term for gestation is reached approximately 280 days after the last menstrual cycle began. Inherent contraction of uterine smooth muscles is inhibited by progesterone. Decreased hormone production by the aging placenta results in a sharp decline in progesterone levels as the end of preganacy nears. The inhibition of uterine contractions is released. *Oxytocin*, produced by the posterior pituitary accentuates uterine smooth muscle contractions. These contractions of the uterine wall **(labor)** expel the fetus through the cervix and vagina.

CHAPTER REVIEW

1. Define homeostasis and its role in survival.

2. Name the four basic tissue types; give examples of each.

3. Describe the structure and function of

   a. three epithelial tissue types
   b. three muscle tissue types

4. Describe the cellular and molecular basis of muscle movement.

5. Describe electrical impulse generation and conduction.

6. Differentiate between the somatic and autonomic nervous systems.

7. Describe the role of hemoglobin in oxygen transport.

8. List the chambers of the heart and describe the function of each.

9. Differentiate between arteries, veins, and capillaries.

10. Describe the location of the lymphatic system. What is its role?

11. Describe the exzymatic digestion and subsequent absorption of

    a. carbohydrates
    b. proteins
    c. fats

12. Describe the structure and function of the nephron.

13. Describe the hormonal controls of the menstrual cycle.

10

GENETICS

Genetics is the study of *heredity* (the process by which traits are passed from parent organism to offspring) and *variation*. It is one of the youngest branches of biology and perhaps the most important from a sociological standpoint. Genetics has had enormous impact on areas such as

agriculture — in the improvement of cultivated plant and animal stocks, and the development of biological controls for pests and disease

medicine — in the understanding, treatment, and prevention of inherited disorders

genetic engineering — in the development of pharmaceuticals, genetic manipulation, and the potential for catastrophic accidents and biological warfare

evolutionary theory — in the contribution of strong supporting evidence

Genetics is also concerned with the cellular, organismal, and environmental factors that affect the mechanisms of inheritance. The study of genetics is approached from several fronts, ranging from the molecular level to the population level.

MENDELIAN GENETICS

The field of genetics is said to have originated with the work of an Austrian monk named Gregor Mendel. Working at about the same time as Darwin was writing *The Origin of Species*, Mendel investigated the principles of inheritance through a series of experiments using the common garden pea. He demonstrated that inherited characteristics are carried by factors that function as discrete units. He further demonstrated that these discrete units are distributed in various ways that result in their being *reassorted* with each new generation. Eventually, these units of inheritance were labeled *genes*.

Chromosomes and Genes

In 1903, Walter S. Sutton first recognized that genes are located in chromosomes. Later studies, particularly in the areas of biochemistry and molecular biology, determined that a **gene** is a sequence of nucleotides (constituting a small part of a DNA molecule) that codes for the production of a single polypeptide chain. Because polypeptides are the backbones of proteins, it was shown that genes code for protein structure and function.

Genes occur in chromosomes in precise sequences. The location of any gene on a chromosome is termed its **locus**. Homologous genes on homologous chromosomes (Chapter 8) occur at corresponding loci. On rare occasions, a spontaneous change in some part of DNA may occur as a result of an environmental factor. This change is called a **mutation** and results in an alteration in the code of instructions for protein synthesis (Chapter 11). Because they can result in the production of defective protein or the cessation of protein synthesis, mutations are usually lethal. On rare occasions, however, mutations can manifest changes in some measurable attribute (in structure or function). If these measurable attributes, termed **traits** or **characters**, are inheritable, then any given gene may have two or more alternative forms. These alternative forms may coexist on homologous chromosomes as **alleles**. Alleles occurring at the same locus on homologous chromosomes are termed a **gene pair** or **allelic pair**. Identical or different alleles may constitute a gene pair. If the alleles are identical, the organism is said to be **homozygous** for that trait; if different, the organism is said to be **heterozygous** for the trait. The genetic makeup of an organism, whether for a single allelic pair or for the total complement of genes, is termed the **genotype**.

Mendelian Laws of Inheritance

Mendel published the results of his work on garden peas in 1866. In his paper, he proposed two fundamental principles, which govern inheritance: the *principle of segregation* and the *principle of independent assortment*.

Principle of Segregation. In sexually reproducing organisms, *during gametogenesis, each member of a pair of hereditary units* (alleles) *segregates to a different gamete*. In the garden pea, for example, the length of the stem is controlled by a single gene for which there are two alleles: tall (T) and dwarf (t). If two pea plants, one homozygous tall (TT) and the other homozygous dwarf (tt), are crossed, the homologous alleles are segregated to different gametes. When fertilization occurs, each parent contributes an allele resulting in a zygote that is heterozygous for stem length (Figure 10.1). The offspring (the F_1 generation) are termed **hybrids** because they result from parents that differ in at least one trait. When organisms are heterozygous for only one trait, they are termed **monohybrids**.

Mendel found that after crossing homozygous tall plants with homozygous dwarf plants, all F_1 offspring were tall plants. He subsequently described the tall form of the trait as **dominant** and the dwarf form as **recessive**. When the F_1 generation were permitted to interbreed at random, only those members of the F_2 generation that were homozygous recessive displayed the dwarf stem

parents: TT   X   tt

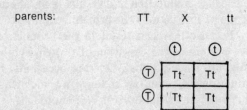

Figure 10.1. A cross of two pea plants, one homozygous dominant for stem length (TT), the other homozygous recessive for the same trait (tt). The Punnett-square method is used to determine the probable ratio of traits appearing in the F_1 offspring. Circles symbolize gametes, the squares offspring. The letters in each depict their genotype. By convention, uppercase letters denote dominant alleles, lowercase letters represent recessive alleles.

parents: TT   X   tt

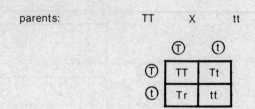

Figure 10.2. Results of a monohybrid cross (involving an F_1 generation of pea plants heterozygous for stem length (Tt).

length (Figure 10.2). Therefore, although there are three different genotypes—homozygous dominant (TT), homozygous recessive (tt), and heterozygous dominant (Tt)—in a dominant-recessive situation there are only two phenotypes—tall-stemmed plants (TT and Tt) and dwarf-stemmed plants (tt). **Phenotype** is the structural and functional manifestations of an organisms's genotype.

Principle of Independent Assortment. Also from his studies of the garden pea, Mendel concluded that *members of each pair of traits* (alleles) *are segregated to gametes independently of any other pair*. A **dihybrid cross** of pea plants involving *two* traits, flower color and length of stem can demonstrate this principle (Figure 10.3). (In the garden pea, flower color is controlled by a dominant allele for purple [P] and a recessive allele for white [p]). If plants heterozygous for both traits are permitted to mate at random, the offspring manifest a phenotypic ratio (9:3:3:1) that reflects the 16 different genotypes expected if the two pairs of alleles assort independently.

Probability

The segregation of homologues to gametes are events that are random in occurrence. A **random event** is an event whose probability of occurrence is not influenced by external forces. **Probability** is defined as the number of times something *will* happen, divided by the number of times it *could* happen. Thus, for a monohybrid cross involving individuals heterozygous for the same trait, two different assortments of alleles to gametes are possible (Figure 10.4). In turn, three different genotypes are possible from four possible combinations of gametes, each genotype having a specific probability of occurrence.

The Laws of Probability. The First Law of Probability states that *the probability of an event occurring will be less than or equal to one*.

$$P_{(a)} = \frac{a}{b} \leqslant 1,$$

parents: PpTt X PpTt

	PT	Pt	pT	pt
PT	PPTT	PPTt	PpTT	PpTt
Pt	PPTt	PPtt	PpTt	Pptt
pT	PpTT	PpTt	ppTT	ppTt
pt	PpTt	Pptt	ppTt	pptt

phenotypes	phenotypic ratio	genotypes
purple-flowered, tall	9	PPTT, 2 PPTt, 2 PpTT, 4 PpTt
purple-flowered, dwarf	3	PPtt, 2 Pptt
white-flowered, tall	3	ppTT, 2 ppTt
white-flowered, dwarf	1	pptt

Figure 10.3. Results of a dihybrid cross between plants heterozygous for two traits, flower color (P = purple flowers, p = white flowers) and length of stem (T = tall plants, t = dwarf plants).

where a = event,
   P = probability, and
   b = total possible outcomes.

Therefore, for an organism that is heterozygous for one trait, the probability that one of its gametes will possess one of the alleles for that trait will be ½ or 50 percent.

  The Second Law of Probability states that *the probability of two separate events occurring simultaneously is equal to the product of their individual probabilities*:

$$P_{(abc)} = P_{(a)} \times P_{(b)} \times P_{(c)}$$

where a, b, and c are three independent events.

For example, the frequency of homozygous tall pea plants (TT) produced from heterozygous parents (Tt) would be as follows:

Parents: Tt × Tt

Gametes: ½T ½t ½T ½t

$$P_{(TT)} = \text{½T} \times \text{½T} = \text{¼ or 25\%}$$

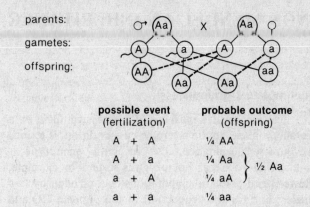

parents:

gametes:

offspring:

possible event (fertilization)	probable outcome (offspring)
A + A	¼ AA
A + a	¼ Aa } ½ Aa
a + A	¼ aA }
a + a	¼ aa

Figure 10.4. Results of a monohybrid cross involving individuals heterozygous for the same trait. (*Note:* ♂ = male and ♀ = female).

A determination of the frequency of heterozygous offspring from the parents would have had to take into consideration that there are two ways this outcome could occur (Figure 10.2). Thus, the resultant frequency would equal ½ or 50 percent.

  Another example might involve determining the frequency of purple-flowered, tall pea plants produced from a cross of plants heterozygous for both traits (PpTt). The first step involves determining the phenotypic frequencies of a monohybrid cross for each individual trait.

Parents: PpTt × PpTt

Pp ×Pp: ¼PP : ½Pp : ¼pp

    ¾ purple-flowered : ¼ white-flowered

Tt × Tt: ¼TT + ½Tt + ¼tt

    ¾ tall : ¼ dwarf

The next step involves applying the Second Law of Probability to determine the frequency of purple-flowered, tall offspring.

$$P_{(purple\text{-}flowered,\ tall)} = P_{(purple\text{-}flowered)} \times P_{(tall)}$$
$$= \text{¾} \times \text{¾} = \text{9/16}$$

Note: The above result is the same as that obtained using the Punnett-square method (Figure 10.3).

NON-MENDELIAN INHERITANCE

Incomplete Dominance

The incomplete masking of a recessive allele by a dominant allele, in the heterozygous condition, is termed **incomplete dominance**. The resultant phenotype is unlike that of either homozygous state. For example, flower petal color in snapdragons is controlled by two alleles: one for the production of red pigment (R) and one for the lack of pigment (white; r). A cross between two homozygous plants (one dominant, the other recessive) results in heterozygous offspring, the phenotype of which is pink petal color (Figure 10.5). One red allele is apparently not enough to code for the pigment production sufficient for red flowers.

Codominance

Codominance is the phenotypic expression of both alleles of a gene locus. For instance, AB-type blood in humans is the result of an A allele (I^A), which codes for the production of A-type antigens on the surface of red blood cells, while the B allele (I^B) codes for the production of B-type antigens. Both types of antigens appear on the same red blood cells, and are therefore codominant.

Multiple Alleles

The ABO blood groups in humans are examples of multiple alleles. **Multiple alleles** is the condition in which more than two alleles exist for a given gene locus. In addition to A and B alleles, an O allele (i) codes for the absence of A and B antigens on the cell membranes of red blood cells. Six possible genotypes result.

Phenotype (blood type)	Genotype
A	I^AI^A, I^Ai
B	I^BI^B, I^Bi
AB	I^AI^B
O	ii

Gene Interaction

Most genes act with others to produce some phenotypic effect. **Epistatic genes** are those which interfere with or prevent the expression of other genes (Figure 10.6). Another example of gene interaction involves **modifier genes**, which alter the phenotypic expression of other genes. For instance, one gene determines the two basic phenotypes of human eye color. Brown-eyed individuals (BB or Bb) have branching pigment cells containing melanin in the front layer of the iris. Blue-eyed individuals (bb) lack melanin in front of the iris. (The blue color of the iris is an effect of the black pigment on the back of the iris seen faintly through the semiopaque tissue layer in front of the iris.) Other genes, however, modify the amount of pigment in front of the iris and its distribution, creating various shades of eye color (black, gray, or green, for example).

Pleiotropy

Some genes have been shown to exert several effects on seemingly unrelated aspects of the phenotype of an organism. This phenomenon is termed **pleiotropy**.

```
parents:        AABB        X        aabb
                (agouti)             (albino)

F₁:                      AaBb
                        (agouti)

                AaBb        X        AaBb

F₂:     ⅛ AaBb    1/16 AAbb    1/16 aabb
        ¼ AaBb    ⅛ Aabb       ⅛ aaBb
        1/16 AABB              1/16 aaBB
        ⅛ AaBB    _____    _____
        9/16 agouti  3/16 black   ¼ albino
```

Figure 10.6. Epistasis. Two independently assorting genes influence coat color in micer. A dominant allele for nonalbinism (A) and a recessive allele for albinism (a) occur at one gene locus. At the other gene locus, a dominant allele (B) produces an agouti (mixture of gray, yellow, and brown) coat, while the recessive allele (b) produces a black coat. Note that the homozygous condition for albinism blocks any expression of the gene for coat color.

```
parents:        RR      X      rr
                red            white

F₁:                   Rr
                      pink

                Rr      X      Rr

F₂:     ¼ RR    ¼ Rr    ¼ Rr    ¼ rr
        red     pink    pink    white
```

Figure 10.5. Incomplete dominance at one gene locus. A cross between red-flowering (RR) and white-flowering (rr) snapdragons yields pink-flowering plants in the first generation. When the F₁ generation is crossed with each other, the F₂ generation has a phenotypic ratio equivalent to its genotypic ratio.

An example of pleiotropy is sickle-cell anemia, a genetic disorder so named because of the distortion human red blood cells undergo due to a variation in the structure of hemoglobin A. Sickling of red blood cells is produced by a single recessive allele (HbS) that codes for a substitution of one amino acid by another in the β chain polypeptide of hemoglobin. A host of other symptoms are also produced (Figure 10.7). In the case of the homozygous condition (HbS/HbS), death occurs. Heterozygotes (HbA/HbS) show few disease symptoms because the one normal allele is fully functional.

Linkage

With the development of models for gene and chromosome structure came the realization that Mendel's Law of Independent Assortment only holds true if different genes occur on different chromosomes. If genes occur on the same chromosome they are said to be **linked**. Chromosomes are sometimes referred to as **linkage groups**. Linked genes become assorted independently only if they are separated during crossing over (Figure 10.8).

Determination of Sex

In most higher organisms where the sexes are separate (mammals, birds, insects, etc.), the chromosomal complement for males and females differs in terms of number or kind. Chromosomes are classified as autosomes and sex chromosomes. **Autosomes** are chromosomes that are the same in number and kind in both sexes; **sex chromosomes** are those that differ in number and kind between sexes. In the cells of some organisms, one sex may possess a chromosome that is absent in the cells of the other sex. For example, the female grasshopper possesses an X chromosome (diploid genotype XX) that is absent in the male (diploid genotype XO). However, two sex chromosomes are generally present in the somatic cells of each sex. The sex chromosome pattern XX (female) and XY (male) is found in most animal species, including *Homo sapiens*. Apparently, the information contained in the combinations of sex chromosomes determines the phenotypic difference between males and females.

Sex-linked Inheritance

Many genes occur on X chromosomes and not on Y chromosomes; therefore, their patterns of inheritance will be different than those on autosomal chromosomes. Specifically, the genes received by a male from one parent will not be homologous to those received from the other parent. Consequently, recessive alleles cannot be masked in the male since the heterozygous condition

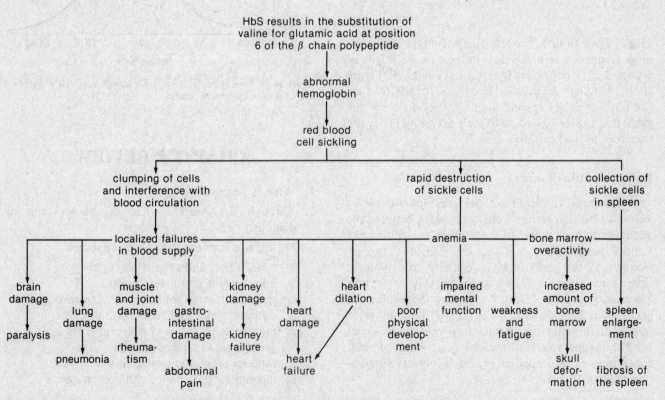

Figure 10.7. Pleiotropic effects of sickle-cell anemia (HbS).

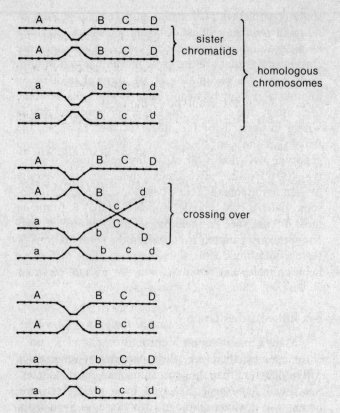

Figure 10.8. Linkage and genetic recombination as a result of crossing over.

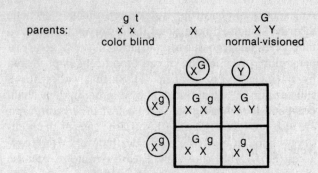

F1:   50% normal-visioned (carrier) females
         50% color blind males

Figure 10.9. Example of sex linkage involving the gene for green weakness, a partial color blindness caused by reduced retinal pigment that is sensitive to the green portion of the visible light spectrum. The allele for normal pigmentation is dominant (G) and that for green weakness is recessive (g). Note that the offspring of a color-blind female and a normal-visioned male would be females who had normal vision and were carriers, and males who were all color blind.

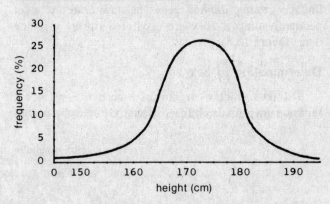

Figure 10.10. An example of polygenic inheritance; height distribution of men in the United States.

cannot exist. In turn, the recessive phenotype will occur more frequently in males than in females. As might be expected, this condition is termed **sex linkage** (Figure 10.9). Although females heterozygous for the trait do not phenotypically express the trait, they do carry one allele that can be passed on to male offspring. They are therefore termed **carriers**.

Polygenic Inheritance

In contrast to pleiotropy, **polygenic inheritance** involves a trait that results from interactions between the alleles of two or more different genes controlling the same inherited characteristics. The individual expressions of all the genes involved combine to produce a single phenotype (as in human height, skin color, hair color, and so forth). Thus, the greater the number of genes involved (polygenes), the greater the number of intermediate genotypes. Traits involving many genes produce phenotypic distributions that, when plotted, conform to a bell-shaped curve, that is, normal distribution (Figure 10.10).

CHAPTER REVIEW

1. What is genetics?
2. Through his research, what was Mendel able to demonstrate?
3. Distinguish between the following terms:
   a. gene, allele, and gene (allelic) pair
   b. homozygous and heterozygous
   c. trait (character), mutation, and hybrid
   d. dominant trait and recessive trait
   e. monohybrid cross and dihybrid cross
   f. genotype and phenotype
   g. random event and probability
   h. incomplete dominance and codominance

i. epistatic genes and modifier genes
j. pleiotropy, linkage, and polygenic inheritance
k. autosomes and sex chromosomes

4. State the First and Second Laws of Probability.

Genetics Problems

The answers to the following problems appear in Appendix B.

1. Find the *genotypic* and *phenotypic ratios* for the following monohybrid crosses involving stem length in pea plants (T = tall plants, t = dwarf plants):

   a. TT × tt
   b. Tt × Tt
   c. TT × Tt
   d. Tt × tt

2. Find the genotypic and phenotypic ratios for the following crosses involving two traits: flower color (P = purple flowers, p = white flowers) and stem length in pea plants:

   a. PPTT × pptt
   b. PPtt × ppTT
   c. PpTT × PPTt
   d. PpTT × PpTt

3. Syndactyly, a condition in which webbing occurs between the fingers, is controlled by a dominant allele (S). The recessive allele (s) determines the development of normal hands. Suppose a man with normal hands marries a woman with syndactyly and their first child is born with normal hands.

   a. What is the genotype of the mother?
   b. What is the probability that their next child will have normal hands?
   c. What is the probability that the next child will be a girl with syndactyly?
   d. What is the probability that the next two births will yield children with normal hands?

4. A woman with O-type blood has a child who also has O-type blood. What blood type could the father *not* have? Why not?

5. A color blind man marries a normal-visioned woman. Their first child is color blind.

   a. What is the genotype of the mother?
   b. What is the probability that the next child will be a boy with normal vision?

6. A color blind man with O-type blood and normal hands marries a woman who is heterozygous for syndactyly and B-type blood, and is a carrier of color blindness. What is the probability that their first child will be a normal-visioned girl with O-type blood and normal hands?

THE MOLECULAR BASIS OF HEREDITY

GENETIC CODE

The genetic basis for the phenotype of any organism resides in the structure of its DNA, which contains the code for protein synthesis. (Heritable traits arise from the actions of structural and enzymatic proteins.) Specifically, the sequence of purine-pyrimidine base pairs (Chapter 3) codes for the assembly of amino acids into polypeptide chains, which are integral to protein structure.

The fundamental unit of heredity is the **gene** — a sequence of nucleotides that codes for the sequence of amino acids of a specific polypeptide chain. Each amino acid is coded for by one or more different **codons**, each of which contains three nucleotides. Four nucleotides, each possessing a different nitrogenous base — **adenine**, **guanine**, **cytosine**, and **thymine**, (or **uracil**) — function as letters for the genetic code (Figure 11.1). As a result, 64 different codons are possible, 61 of which code for the 20 naturally occurring amino acids (Figure 11.2). The other three codons function as "stop" messages in the transcription of the DNA code into the structure of RNA.

DNA REPLICATION

In cell reproduction and gametogenesis, genetic information, in addition to cytoplasmic material, is passed on to daughter cells (Chapter 8). DNA replication precedes both processes, providing sufficient copies of genetic information for the daughter cells.

In the duplication of DNA, the double helix first unwinds and the weak hydrogen bonds linking the base pairs together (Chapter 3) are broken. As a result, the two strands of DNA separate in a zipper-like fashion, each becoming a template for the formation of a complementary strand (Figure 11.3). The assembly of new strands of DNA results from the attraction of free nucleotides to complementary bonding sites on the original DNA strands. The sequencing of the nucleotides is determined by a mechanism termed **base pairing**, in which the purine adenine (A) can only bond with the pyrimidine thymine (T), and guanine only with cytosine (C). The synthesis of new DNA strands is accomplished under the direction of the enzyme **DNA polymerase**.

RNA AND PROTEIN SYNTHESIS

Ribonucleic acid differs from deoxyribonucleic acid in three ways:

1. Unlike most DNA, RNA is composed of a single chain of nucleotides.
2. The pentose sugar in RNA is *ribose*, not deoxyribose (Figure 11.4).
3. In the place of thymine, RNA has *uracil* (Figure 11.1), which behaves the same as thymine chemically.

There are three kinds of RNA: **messenger RNA** (mRNA), **transfer RNA** (tRNA), and **ribosomal RNA** (rRNA).

Transcription and Translation

Protein synthesis begins with the transcription of the encoded message in DNA into the structure of RNA. **Transcription** is the assembly of nucleotides, directed by

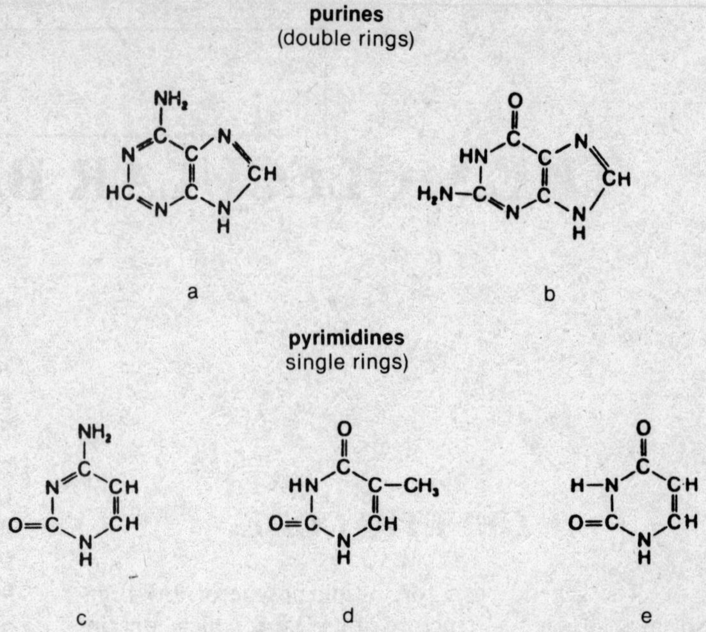

purines
(double rings)

Figure 11.1. Nitrogenous bases: (a) adenine, (b) guanine, (c) cytosine, (d) thymine and (e) uracil. Nucleotides of DNA contain adenine, guanine, cytosine, and thymine. Thymine is replaced by uracil in RNA structure.

a

b

pyrimidines
single rings)

c

d

e

First letter		Second letter				Third letter
		U	C	A	G	
U		phenylalanine	serine	tyrosine	cysteine	U
		phenylalanine	serine	tyrosine	cysteine	C
		leucine	serine	stop	stop	A
		leucine	serine	stop	tryptophan	G
C		leucine	proline	histidine	arginine	U
		leucine	proline	histidine	arginine	C
		leucine	proline	glutamine	arginine	A
		leucine	proline	glutamine	arginine	G
A		isoleucine	threonine	asparagine	serine	U
		isoleucine	threonine	asparagine	serine	C
		isoleucine	threonine	lysine	arginine	A
		isoleucine	threonine	lysine	arginine	G
G		valine	alanine	aspartate	glycine	U
		valine	alanine	aspartate	glycine	C
		valine	alanine	glutamate	glycine	A
		valine	alanine	glutamate	glycine	G

Figure 11.2. The genetic code. The 20 naturally occurring amino acids are each coded for by one or more nucleotide triplets (codons). In addition, three codons function as "stop" messages. The first nucleotide of any codon is designated by the left column, the second by the middle columns, and the third by the right column. For example, phenylalanine is coded for by the codons UUU and UUC.

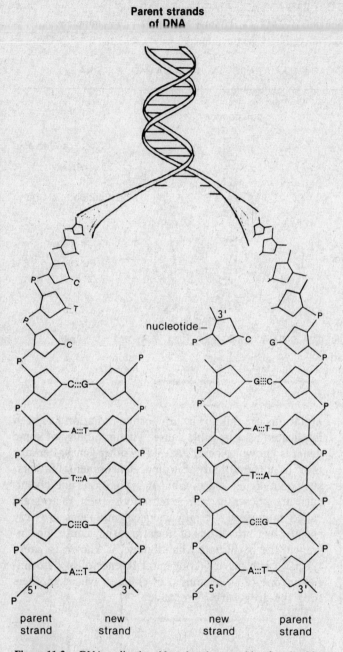

Parent strands of DNA

nucleotide —

parent strand | new strand | new strand | parent strand

Figure 11.3. DNA replication. Note that the assembly of nucleotides into a new strand proceeds in the 3′→5′ direction. (By convention, the carbon atoms of the sugar are numbered to enable identification of specific regions of the molecule, which in turn facilitates orientation to the nucleotide or nucleic acid of which the sugar is a part.)

a   b

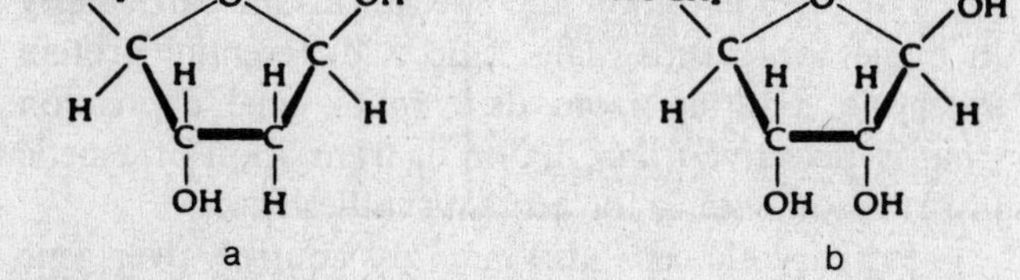

Figure 11.4. Deoxyribose (a) and ribose (b) are five-carbon sugars that differ only in that deoxyribose has one less oxygen atom.

the enzyme **RNA polymerase**, into a sequence complementary to that of the DNA (Figure 11.5). (Presently, it is believed that only one strand of DNA functions as the template for RNA synthesis.) The genes coding for structural and enzymatic proteins are transcribed into individual mRNA units. The mRNA serves as a template for the assembly of amino acids into a polypeptide chain.

Completed mRNA moves from the nucleoplasm, through the nuclear envelope, and into the cytoplasm where one or more ribosomes become attached to it. Ribosomes move unidirectionally along mRNA in the 5′→3′ direction (Figure 11.6). (By convention, the carbon atoms of the sugar are numbered to enable identification of specific regions of the molecule, which in turn facilitates orientation to the nucleiotide or nucleic acid of which the sugar is a part.) Ribosomes serve as the sites of interaction between mRNA and tRNA. (Ribosomal RNA is thought to provide a platform for the assembly and positioning of ribosomal proteins involved in protein synthesis.) Amino acids are transported to sites of protein synthesis by tRNA molecules. Amino acids specific to each tRNA attach at one end of the nucleic acid. A triplet of bases located at the other end form an **anticodon** (Figure 11.6). The anticodon is a template recognition site that is complementary to a specific mRNA codon with which it can subsequently hydrogen-bond. The specificity of codon-anticodon interactions is precise between their first two base pairs, but the third base pair is less precise. Thus, some tRNA molecules can "recognize" (hydrogen-bond to) more than one codon. The mutual affinity of complementary base triplets brings each amino acid into proper sequence with others of the growing polypeptide chain (Figure 11.6). Enzymes thought to be present in the large subunit of ribosomes mediate the formation of polypeptide bonds as amino acids are added to the chain. Both the amino acid and the mRNA then release the tRNA, which can subsequently attach to another amino acid of the same type. Translation of the nucleotide code of mRNA into an amino acid sequence is completed when ribosomes reach the end of the mRNA, where a "stop" codon occurs to which a tRNA will not bond.

IMPLICATIONS OF MUTATION

A **mutation** is any heritable change in the DNA of offspring that is absent from the genetic complement of the parent generation (Chapter 10). The change may apply to the kind, structure, sequence, or number of the component parts of DNA. Mutations have been classified into two major categories: point mutations and

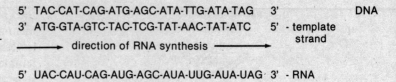

Figure 11.5. Transcription: the transfer of genetic information from DNA to RNA.

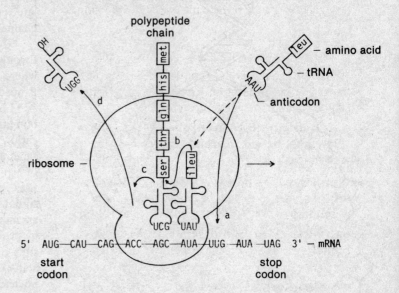

Figure 11.6. Translation: the conversion of genetic information into the amino acid sequence of a polypeptide chain. The steps illustrated above include: (a) attachment of amino acid-carrying tRNA to a complementary codon, (b) peptide bond formation, (c) breaking of amino acid-tRNA bond, and (d) ejection of tRNA from the ribosome.

chromosomal mutations. **Point mutations** (also termed **gene mutations**) involve alterations of single genes and are not detectable cytologically. In contrast, **chromosomal mutations** involve gross changes in the kind, structure, or number of DNA molecules.

Mutations are normally very rare. For instance, in DNA replication, it has been estimated that for every 9 to 10 billion base pairs assembled, one mistake is made. When mutations do occur, they are usually harmful. Because of the high degree of coordination within living systems, and their precise adaptation to a given environment, even the slightest change in genetic information can be lethal. For instance, sickle cell anemia (Chapter 10) is thought to have originated in West Africa as a change in just one base pair out of about 450 coding for the β chains of human hemoglobin. The result was a change in just one amino acid out of about 150 composing the β chains. In the homozygous condition, sickle cell anemia usually results in death during young adulthood. However, in the heterozygous condition, the mutation does not usually cause death and is actually beneficial to those living in malaria infested regions (e.g., West Africa), by providing a resistance to malaria. Thus, mutation is an important source of *variation* on which adaptation and, therefore, evolution depend.

Many causes of mutations have been identified or implicated. For instance, the causes of spontaneous mutations are unknown, but cosmic radiation has been implicated as a possible cause. The mutability of some genes is known to be influenced by other genes. Perhaps most important, many abnormal environmental factors, such as ionizing and nonionizing radiations and chemicals, are known to *induce* mutations; these factors are called **mutagens**. Certain chemical mutagens are very specific as to the kinds of mutations they cause. For instance, the barbituate, thalidomide, is known to cause specific birth defects (congenital absence of or deformities in one or more limbs) in children whose mothers took the drug during pregnancy.

GENE EXPRESSION

Normally, all somatic cells of multicellular organisms, with a few exceptions (e.g., red blood cells), have the same genetic information. Yet, they show high degrees of differentiation and specialization, particularly in higher organisms. This reflects differential protein synthesis resulting from differential gene expression (that is, selective transcription or translation of genetic information coding for protein synthesis).

Prokaryotic cells also manifest controls over gene expression. For example, bacteria grown on a medium completely lacking the sugar lactose produce no lactose-degrading enzymes. However, if the same bacteria are abruptly switched to a medium in which lactose is their only energy source, they quickly begin to produce the necessary enzymes to catabolize lactose. Thus, the bacteria possessed the genes for such enzymes, but the genes were expressed only when the need arose.

Five general mechanisms control gene expression:

transcriptional controls — control the access of enzymes of transcription to genes coding for structural and enzymatic proteins

post transcriptional controls — edit RNA transcripts before reaching ribosomes for translation

translational controls — determine which mRNA molecules will be translated at any given time

post translational controls — modify the RNA products of genes or the proteins for which they code

enzymatic controls — affect the activity of enzymes directly or indirectly involved in protein synthesis

CHAPTER REVIEW

1. What is the fundamental unit of heredity? How is it involved in the molecular basis of protein synthesis?

2. Describe the process of DNA replication.

3. Compare and contrast DNA and RNA structure.

4. Protein synthesis involves two steps: transcription and translation. Describe the processes involved in each.

5. The following represents the transcription and translation of information encoded in the structure of DNA. Fill in the blanks with the appropriate complimentary nucleotide triplets.

$5'$ T__C -__AT-__AG-AGC-_____-_____ $3'$ template $\Big\}$ DNA
$3'$ ATG-__TA-G__C-TCG-_____-A____ $5'$ strand

———→ direction of transcription ———→

$5'$ UAC-CAU-CA__-_____-AUU-__A__ $3'$ mRNA

———→ direction or translation ———→

A__G __UA _____ UCG _____ ___G tRNA units

6. Define the following terms: mutation, point (gene) mutation, chromosomal mutation, and mutagen.

7. Contrast the five general kinds of controls over gene expression.

THE PRINCIPLES OF EVOLUTION

Evolution is a process; it obeys the fundamental laws of nature. It is *not* a plan with intent and purpose. Environmental conditions *of the present*, not future goals, determine the direction of evolutionary change. As a result of evolution, organisms are able to survive in a changing environment. Evolution operates at the population level; individuals survive, populations evolve.

EARLY EVOLUTIONARY THEORY

Ladder of Life

Aristotle envisioned the *Scala Naturae* (ladder of nature) in which all living organisms exist in a hierarchy. The simplest creatures inhabit the bottom rungs while humans occupy the top. He also developed the theory of **spontaneous generation**: living organisms arise from nonliving and unrelated components. Reviving Aristotelian concepts, European Renaissance scholars also ordered living things in a "ladder of life"; however, in keeping with Judeo-Christian teachings, they believed in **original creation**, that living things were products of divine intervention. Neither explanation postulated a common historical relationship among species.

Preevolutionary Ideas

The 18th century French scientist, Georges de Buffon, first proposed that species change throughout time. Buffon proposed that in addition to those species created at the beginning of the world, forces of Nature shaped new species by improvement and degeneration.

The French clergy eventually forced him to retract these ideas.

Eighteenth-century geologists played an important role in the development of evolutionary theory. The theory of **uniformitarianism** stated that the Earth had a long history and that slowly, during time, gradual processes shaped the present-day planet. *Strata* (layers of rock) were discovered to contain characteristic groups of fossil specimens. Fossils were no longer studied as accidents of nature, but as a record of the link between the history of the Earth and the history of living organisms.

Georges Cuvier's **catastrophe theory** silenced these early evolutionary ideas. He postulated a series of catastrophes to explain the fossil record and extinction of species. After each violent event of nature, those species that remained alive repopulated the Earth. Time and nature acted only to eliminate, not create species.

The first systematic theory of evolution was proposed in 1801 by Jean Baptiste Lamarck. He proposed that all species shared a common ancestry. Lamarck noted that older rocks contained only simple life forms. He interpreted this to mean that simple life forms became higher life forms through a progression of events. Two forces drove this Lamarckian view of evolution: *the inheritance of acquired characteristics and an unconscious striving toward greater complexity.* The first concept was an early attempt to explain how animals changed. Lamarck theorized that body parts became stronger or weaker through use or disuse. These changes were then transmitted by the parent to its offspring. (Experiments have now shown the *combined* influence of environmental and genetic information on the structure of body parts. Acquired characteristics, or developmental changes, of themselves, are not inheritable characteristics.) The second concept — the striving towards greater complexity — predicted that every organism lower on the scale of life was on the way to becoming *Homo sapiens*. Those diverted in an unfavorable environment never proceed past some lower life form. The metaphysical force, the *universal will* was present nonetheless.

Many current opponents of evolutionary theory are set at a Lamarckian reference point. However these vague Lamarckian concepts are now disregarded by modern evolutionary theory.

DARWINIAN EVOLUTIONARY THEORY

Two of Charles Darwin's contemporaries greatly influenced his thinking and his theories. Charles Lyell, a geologist, opposed the catastrophe theory. He proposed continuous geologic changes throughout Earth's history, changes not obvious in a single lifetime. Darwin reasoned that these same physical forces act over the same long period of time on living organisms as well.

The second influence came from the writings of Reverend Thomas Malthus. His *Essay on the Principle of Population* (1798) warned that increased demand for food would outweigh the world food supply; the increasing human population would soon run out of room. Noting that population size tended to remain constant, Darwin concluded that populations were held in check by food supply and other factors.

The Origin of Species

Twenty years after his voyage with the *Beagle*, Darwin pondered the many observations he had made. His careful observations as a naturalist and a geologist are combined in his publication, *The Origin of Species* (1859). In this treatise, Darwin commented on the variations observed in all species of organisms. These variations serve to distinguish one species from another, as well as one individual from another. Darwin proposed that *ever-changing environmental conditions operated on these variations, favoring and stabilizing some variations while discouraging and eliminating others*. His theory of **natural selection** states that varied individuals within a population must compete for limited resources (food and shelter, for example). Individuals with certain variations are better adapted to the changing environment. They survive and reproduce, passing those variations to the next generation. If the same environmental pressures exist, the new generation will also be better adapted to their environment, be more likely to survive, and leave offspring. *Natural selection, then, is a theory of differential reproduction based on differential survival*. The pressures of environmental conditions drive natural selection.

According to Darwin, variations within a population occur at random and are not the result of a "striving toward complexity" as Lamarck had theorized. The variation itself is neutral. The forces of natural selection place the positive (survival) or negative (death) value on the variation. If a variation has no negative effect on an individual, the individual has a greater chance of leaving surviving offspring.

Hypotheses of Evolutionary Theory

Three hypotheses are involved in the theory of evolution:

1. There are more individuals produced than reach reproductive age.
2. Variations occur randomly within a population and some variations can be passed on to the next generation.
3. Survival is determined by fitness, not chance. Differential survival determines the contribution made to the next generation.

MODERN EVOLUTIONARY THEORY

Darwin was unable to explain the source of the variations he observed among individuals. Nor was he able to explain how this variation was transmitted to an individual's offspring. The rediscovery of Mendelian genetics and the elucidation of DNA's role in heredity solved these ambiguities. Modern evolutionary theory states:

1. Mutation, change in the genetic makeup, occurs continually within a population.
2. Natural selection acts on these mutations (genetic variations).
3. Once a selective change has occurred, a degree of isolation is needed to establish the genetic change within the gene pool of the population. Modern studies of these evolutionary theories are carried out by population geneticists.

Population Genetics

Changes during the lifetime of an individual organism are **developmental changes**, whereas **evolutionary changes** are established over many generations.

As stated earlier, natural selection acts upon the individual's variations; individuals survive, populations evolve.

A **population** is a group of interbreeding organisms. All genetic information within a population is termed the **gene pool**. Only a small part of the gene pool resides within each individual. In a randomly breeding population, all individuals contribute to the gene pool. The genetic contributions of the entire population affects the population's ability to change and survive over time. The gene pool of small, isolated populations frequently carries less variation.

Variation. The genetic contribution of each surviving individual is crucial. No two individuals in a population are identical (except monozygotic twins). Each individual holds a unique portion of the gene pool — a unique set of genes termed a **genotype**. These variations manifest themselves as differences in appearance, **phenotypes**. Often several genes interact to produce a single phenotype (Chapter 10).

Genotypic variations within a population originate by: random changes of the DNA (**mutation**), rearrangement of genetic material (**recombination**), and the movement of new individuals into the population (**migration**).

Mutations occur in several ways (Chapter 11). Single base-pair changes in the DNA are **point mutations**. Changes in chromosome number result from loss or gain of one chromosome (**aneuploidy**) or duplication of an entire set of chromosomes (**polyploidy**). Chromosome structural changes occur when portions of the DNA are: repeated (**duplication**), left out (**deletion**), broken and reinserted in abnormal ways (**translocation**), or rearranged in opposite orientation (**inversion**).

Recombination of genetic material occurs during sexual reproduction (Chapter 8). Crossing over during meiotic prophase forms entirely new chromatids. New chromosomal combinations result from the pairing of maternal and paternal chromosomes.

Populations adapt to changing environmental conditions and their variations become established within the gene pool. Migrants from another population carry these variations within their portion of the gene pool. If the migrants breed, they contribute new genetic alleles. This is termed **gene flow**. Migration is a significant source of variation only when it involves individuals from two relatively isolated populations.

Selective Pressure. Natural selection is the physical force that systematically results in the differential reproduction. **Artificial selection** occurs when humans select breeding stock on the basis of desired characteristics. The same biological principles are involved in both.

Elimination of extreme forms in a population, favoring the average phenotype is called **stabilizing selection**. **Directional selection** allows individuals at one extreme to survive. **Disruptive selection** discriminates against those individuals with the average phenotype, producing the extreme forms of a species. Most selection is toward stabilization. Variation is conserved and at the same time is directed toward an optimum range.

Maintaining and Restraining Variation. Once a genetic variation has surfaced within a population, forces are at work to keep it or to eliminate it. The push to preserve genetic variety is based on the fact that it is advantageous for a population to maintain many genes in the gene pool — even potentially harmful genes. Environmental changes, or migration, for example, may increase the adaptive value of genotypic variations. A variation previously useless (or potentially harmful) may gain importance once a change has taken place.

Another factor that preserves genetic variety in a population is **heteroses**, adaptive superiority of the heterozygote. The heterozygote is not identical to the homozygous dominant, as was once assumed. The heterozygote carries the alleles found in both homozygotes and is consequently better adapted in certain conditions.

Random changes and accidents (rather than selective pressure) may *establish* genetic variations within a population, causing **genetic drift**. In small populations that have little or no migration, the impact of the variation is significant. Geographic isolation contributes to genetic drift by increasing the frequency of inbreeding.

Castle-Hardy-Weinberg Principle. A model has been proposed to explain the constancy of population gene frequencies under a set of ideal conditions. The **Castle-Hardy-Weinberg Principle** states that the relative frequency of gene alleles remains constant from generation to generation *if*:

- selection of mates is random
- the mutation rate is insignificant
- all genotypes are equally fit and leave equal numbers of surviving offspring
- no immigration (into) or emigration (out of) the population occurs
- random sampling errors do not occur because of small population size

The random association of gametes during fertilization is assumed to be equivalent to the random selection of mates. The principle also assumes that a single locus is responsible for the phenotype. Mathematically,

where *p* is the frequency of the dominant allele, *q* is the frequency of the recessive allele, and since there are only two alleles, p + q = 1. The Castle-Hardy-Weinberg Principle is stated

$$p^2 + 2pq + q^2 = 1$$

where p^2 = the frequency of the homozygous dominant individual,

$2pq$ = the frequency of the heterozygous individual,

and q^2 = the frequency of the homozygous recessive individual.

The Castle-Hardy-Weinberg distribution is used to determine the frequency of the heterozygote within a population where the heterozygous individual is indistinguishable from the homozygous dominant. If the frequency that the homozygous recessive individual appears within the population can be determined through sampling, the frequency of the heterozygote can be calculated. Using these calculations, the frequency of the "carrier" of a specific genetic condition can be determined.

In actuality, the principle is a *model* of a hypothetical population. A given population can be measured against this model to determine the extent of change in gene frequencies. Evolution can be defined as the process of change within genetic constitution of a population. By comparing the gene frequencies of a current population with the hypothetical population, it is possible to determine whether the current population is evolving.

Patterns of Evolution. Populations can evolve into separate species. This process of **speciation** occurs chiefly through geographic isolation, which in turn leads to *genetic isolation*. Geographically isolated populations adapt to their own environment and genetic variations become established. Eventually the populations are unable to interbreed (produce viable offspring) even if the geographic barrier is removed. At that time, separate species exist.

The process of producing many different forms from one ancestral form is called **divergent evolution**. Differences result from adaptations made to specific environments. An example of divergent evolution is the many body forms of mammals. **Convergent evolution** occurs when species with differing ancestry occupy similar environments with similar selective pressures. Adaptive changes produce similar variations within each species and the species may share similar phenotypic characteristics. Convergent evolution is thought to

be responsible for the body form resemblance between fish and aquatic mammals (dolphins) *and* birds and flying mammals (bats).

EVIDENCE OF EVOLUTION

Evidence of evolution can be found in the following sources: the fossil record, direct observations, comparison of homologous and analogous structures, comparitive embryonic development, and biochemical studies. A discussion of each source follows.

Fossils serve as indirect evidence of evolution. Remains of plants and animals show that changes have indeed taken place over time. Unfortunately, the fossil record is incomplete. Often, only parts of the fossils are found and *reconstruction* — comparing fossilized parts with similar living forms — is necessary.

Direct Observation. Direct observation of living organisms provides evidence supporting the theory of evolution. During his famous sea voyage in 1831, Darwin observed 14 species of finches, each similar yet having distinct differences in beak structure. Darwin explained the differences by correlating the diet and habitat of each population. However, these differences in beak structure could have resulted from many genetic changes occurring in isolated populations over long periods of time. *Observations do not prove theories. Theories serve to explain observations.*

Homologous Structures. Homologous structures (bird wings, alligator forelimbs, whale flippers, and human arms, for example) have similar structure but different functions. Darwin suggested that homologous organs and body parts exist as a result of divergent evolution. Homologous structures do not *prove* evolution, but the theory of evolution explains observations of homologous structures. A continual increase in minor genetic variations within a population results in the divergent evolution of populations possessing homologous structures.

Analogous Structures. Analogous structures are similar in function but different in structure. The link between organisms possessing analogous structures and a common ancestor is remote in time. For example, the wings of a butterfly and a bat are analogous, as are the eyes of a squid and a mammal. But detailed study of the development, tissue structure, and organization shows that these body parts are very different.

Embryonic Development. Vertebrate embryos are remarkably similar to each other in their early embryonic development. However, as development proceeds, they become progressively different. This similar pattern in embryonic development supports the theory of evolution. The genetic differences that appear during embyronic development are collections of genetic changes accumulated step by step over long periods of time as species evolved from a common ancestor.

Biochemical Studies. Comparative biochemistry measures genetic similarity. Phenotype is dependent upon genotype. If two groups of organisms appear similar, their DNA must also be similar. It is now possible to distinguish single amino acid changes in proteins and single base changes in DNA. The differences are interpreted as the amount of evolutionary change that has taken place. For example, Cytochrome c, an electron carrier protein, has been extensively studied in over 50 species. Of approximately 110 amino acids, 27 are invariable from species to species. Several other amino acids are only infrequently replaced and then by only one other amino acid. A comparison of horse and yeast Cytochrome c yields 48 differences, whereas only 2 differences occur in the Cytochrome c of chickens and ducks.

Combined information from comparative anatomy, embryology, and comparative biochemistry has led to the construction of detailed phylogenetic trees that demonstrate scales of relatedness among organisms.

MICROEVOLUTION

Microevolution, the change of small portions of an organism's genetic information, is taking place today. Direct observation of organisms that have short life spans can take place in the laboratory where many generations can be studied. For example, billions of bacteria can be grown in the laboratory. If they are grown in medium containing the antibiotic drug, tetracycline, randomly-occurring mutations will make some cells resistant to tetracycline. Due to the selective pressure of tetracycline in the environment, all other cells die and the tetracycline-resistant cells survive. Growth and division of the resistant type eventually forms a new population of cells. Resistance of insects to insecticides, plants to herbicides, and bacteria to drugs, are examples of microevolution.

CLASSIFICATION OF LIFE

Evolutionary theory provides the basis for the logical grouping of living organisms. Demonstrable resemblances are the result of common origins; demonstrable differences are the result of the accumulation of genetic changes over millenia.

Taxonomy and Systematics

Taxonomy is the science of identification, the naming and classification of organisms. Each organism is classified into eight categories (taxon): Kingdom, Phylum, Class, Order, Family, Genus, Species, Subspecies. The first taxon is the most general; the last is the most specific.

Systematics studies the evolutionary relationship between organisms. The systematist constructs phylogenetic trees to depict these relationships. The following assumptions form the basis for systematic biology:

1. Taxon characteristics and similarities can be identified; closely related categories share morphological and biochemical similarities.

2. Genetic changes lost in evolution tend not to be reintroduced into a population. Therefore, similarity decreases as time-since-common-ancestry increases.

3. Homologous structures resulting from common genetic ancestry can be differentiated from analogous structures resulting from similar environmental pressures.

Using these assumptions, the systematist forms predictive hypotheses. These, in turn, form the basis for biochemical, genetic, physiological, and anatomical studies of evolution.

The Five Kingdom Scheme

There is no such thing as *the* correct system of classification; classification is always open to discussion, disagreement, and rearrangement based on the scientific method.

Folklore and tradition formed the basis for the Two Kingdom Scheme in which every living organism was either plant or animal. In this scheme, Euglena was classified as an animallike plant by botanists and as a plantlike animal by zoologists. **Euglena** are photosynthetic organisms (the characteristic designated as plantlike), but their gullet structure was thought to function in the ingestion of food (the characteristic designated as animallike). Fungi were classified as plants, even though they are not photosynthetic; yet, they do not move around and ingest solid food. Bacteria were placed in the Plant Kingdom because they have rigid cell walls and do not ingest food. Clearly, this classification scheme presented many problems.

In 1969, R. H. Whittaker proposed the Five Kingdom Scheme for classifying organisms (Table 12.1). This method of classification identified not only the Plant and Animal Kingdoms, but also differentiated the Monerans, the Protists, and the Fungi. The Five Kingdom scheme is also provisionary; it does not account for the Viruses (which some would argue are nonliving).

Table 12.1. Summary of Whittaker's Five Kingdom Scheme.

	Monera	Protists	Fungi	Plantae	Animalia
Cell type	prokaryotic	eukaroytic	eukaroytic	eukaryotic	eukaryotic
Body form	single-celled	mostly single-celled	limited multicellularity (coenocytic)	single-celled and multicelled species	multicelled
Cell differentiation	absent	absent	limited to certain stages in life	present	present
Cell wall materials	present in all (peptidoglycan)	present in some (various types)	present in all (chiton and noncellulose polysaccharides)	present in all (cellulose)	absent

CHAPTER REVIEW

1. Define evolution.
2. Outline the theories of Aristotle, original creation, Cuvier, and Lamarck.
3. Define Darwin's theory of natural selection.
4. List the three hypotheses of evolutionary theory.
5. Define population and gene pool.
6. What are the possible sources of variation within a population?
7. What is the Castle-Hardy-Weinberg Principle? How is it used to determine population gene frequencies?
8. What forces act in a population to maintain and to eliminate genetic variation.
9. Differentiate between divergent and convergent evolution and provide examples of each.
10. Outline six types of scientific observations that support the Principle of Evolution.
11. Outline the key features used to differentiate organisms in the Five Kingdom Scheme.

KINGDOM MONERA (PROKARYOTAE)

All prokaryotic single-celled organisms are placed within the Kingdom Monera.

CELL STRUCTURE

The major difference between prokaryotic and eukaryotic cells is that all prokaryotic cells lack a nucleus.

The plasma membrane of prokaryotes is called the unit membrane (similar in basic structure to the cell membrane described in Chapter 4). The **unit membrane** is a bimolecular leaflet of phospholipid with embedded proteins. It is a semipermeable membrane acting as the agent for selective transport in and out of the cell and serves as the site for respiration and photosynthesis. In eukaryotic cells, these functions are localized within specialized organelles. Prokaryotic organisms, lacking membrane-bound organelles, carry out these functions on internal invaginations of the plasma membrane. This increases the surface area, effectively increasing the efficiency of the processes.

The cell wall is vitally important to prokaryotes. It provides shape as well as osmotic protection. Lacking a contractile vacuole to pump out the water flow from hypotonic environments, the rigid cell wall prevents cell lysis. **Peptidoglycan**, a polymer unique to prokaryotic cells, is found in the cell wall. The monomer consists of sugar derivatives attached to a series of amino acids. The amino acids are crosslinked to amino acids of another peptidoglycan monomer. Penicillin, an antibacterial drug, inhibits the enzyme that crosslinks the peptidoglycan monomers and can cause cell lysis in certain bacteria.

The **ribosomes** of prokaryotes are distinct from eukaryotic ribosomes in that they are smaller, lighter in weight, and do not attach to endoplasmic reticulum while reading messenger RNA and synthesizing proteins.

The DNA in bacterial cells is packed into a region known as the **nuclear region** or **nucleoid**. The DNA molecule is called a chromosome, although it does not contain proteins as does a eukaryotic chromosomes. The **cytoplasm**, everything outside the nuclear region, is immobile. Prokaryotes do not exhibit amoeboid movement.

Structures specialized for motility, **flagella**, are found in some prokaryotes. A prokaryotic flagellum lacks the eukaryotic arrangement of microtubules. Rather, it is composed of flagellin protein subunits. The flagellum attaches to the outside of the cell through the cell wall and plasma membrane. Movement is thought to occur by flagellar rotation.

Additional surface structures are **fimbriae** and **pili**. They are not involved in motility. Fimbriae are short and numerous. They confer "stickiness" to the cell and are involved in cellular attachment. Pili are structurally similar to fimbriae but are longer and fewer in number. They appear to be involved in the mating (conjugation) process.[1]

Cell Size and Shape

Prokaryotic cells are smaller than eukaryotic cells. Most prokaryotes range in size from 0.5–1.5 microns (micrometers) in width by 1–3 microns in length.[2] The smallest free-living organism is approximately 0.2 micron in diameter. The largest prokaryotes are some of the cyanobacteria, 5 microns by 50–100 microns in size.

Prokaryotes are generally classified into one of three shapes. Sperhical, nonmotile cells are known as **cocci**, and exist as individual cells, or in pairs, chains, or clusters. Motile, rod-shaped cells are: the **bacillus**, the **spirillum** (spiral rod-shaped cell), and the **vibrio** (rod-shaped cell with a single curve). The third prokaryotic shape is **coenocytic** — branching tubes (hyphae) form a filamentous mat (mycelium), similar to molds.

BIOCHEMICAL DIVERSITY

The Kingdom Monera possesses the full range of metabolic and energy production schemes as presented in Chapter 6. Every known naturally occurring organic substance can be metabolized by some prokaryotic species. This diversity has made the prokaryotic microorganisms efficient exploiters within the biosphere.

Chemoheterotrophs and Chemoautotrophs

The chemoheterotrophs are the most widespread in terms of numbers of species and variety of niche development. **Chemoheterotrophs** obtain energy from breakdown of organic compounds. Different species use different substances as reducible substrates. **Aerobic chemoheterotrophs** use atmospheric oxygen. The **denitrifiers** oxidize nitrates to form atmospheric nitrogen. The **sulfate-reducers** living in mud and rotting materials form hydrogen sulfide, the characteristic "rotten egg" odor. **Fermentative bacteria** use organic compounds as both electron donor for energy production and final electron acceptor. Their carbon sources are varied and include woody substances, cellulose, pectin, agar, rubber, leather, organo-metallic compounds, hydrocarbon gases, and the keratin of hair, horns, and nails.

Chemoautotrophs are totally unique in nature — they can grow in the absence of light on a completely inorganic nutrient base. Carbon dioxide provides their carbon source. Energy sources are hydrogen, ammonia, and sulfides.

Photoheterotrophs and Photoautotrophs

Photoheterotrophs obtain their energy from light and their carbon from organic compounds. This combination (energy/metabolism schema) is also unique to the Monerans. These organisms use organic compounds as the electron source during photosynthesis. They do not produce oxygen; they function as phototrophs only in an anaerobic environment.

Photoautotrophs use reduced inorganic compounds as an electron source during photosynthesis. These compounds include carbon dioxide, hydrogen gas, sulfur compounds, and water. Cyanobacteria use water during photosynthesis and are oxygen-evolving phototrophs.

Facultatives

Individual species of bacteria show remarkable flexibility; they may belong to more than one of the major energy/metabolism categories. **Facultative organisms** are able to produce energy and continue metabolism under different sets of environmental conditions. Facultative anaerobes, for example, use oxygen when it is present yet continue to function in its absence. Some species of chemoautotrophs make use of organic nutrients when they are available. In the presence of oxygen some photoheterotrophs change to aerobic chemoheterotrophic functioning.

TAXONOMY

Taxonomy and classification of Monerans is based on a variety of characteristics.[3] Among the factors used to classify these organisms are: *biochemical data* (nucleic acid sequence homology and metabolic pathways for energy production), *growth requirements* (temperature, pH optima, and relationship to oxygen — aerobic/anaerobic), and *cell structure* (cell wall constituents, method of motion, cell inclusions, storage products, pigments, and endospore production).

Phylum Archaebacteria

Archaebacteria include the **methanogens** (strict anaerobes that reduce carbon dioxide to methane), the **halophilic bacteria** (grow at saturating concentrations of sodium chloride), and the **thermoacidophiles** (use sulfuric acid as a proton source, and grow at 80 degrees Celsius, pH 2). Archaebacteria exhibit several biochemical differences from the other Monerans. Ribosomal RNA sequences are different from both eukaryotes and other prokaryotes. Ether linkages between membrane lipids are found only in archaebacteria; all other cells exhibit ester linkages. Two compounds found in eubacterial and cyanobacterial cell walls, muramic acid and peptidoglycan, are lacking in archaebacterial cell walls.

Phylum Eubacteria

Eubacteria or true bacteria grow as single cells. Some species, however, grow in colonies. Most eubacterial cocci are saprophytes, that is, use nonliving food sources. The bacillus-shaped eubacteria are important decomposers and nitrogen fixers. They are also an important source of antibiotics. Most notable is the bacterium, *Escherichia coli*, found in the human intestinal tract. *E. coli* functions in the final digestive process in the large intestines and manufactures and secretes vitamin K and biotin.

Some species of eubacteria are disease causing. Among these are the bacteria that cause strep throat (Streptococcus), staph infections (Staphylococcus), tuberculosis, typhoid, tetanus, meningitis, whooping cough, dysentery, diphtheria, botulism, and anthrax.

Phylum Myxobacteria

The **myxobacteria** are the gliding bacteria; they creep or glide along a secreted slime layer. Their reproduction technique is similar to slime molds — individual cells converge into a mass of cells and a multicellular fruiting body develops. A cyst with reproductive spores sits atop a stalk. Due to this apparent cellular specialization, they are considered multicellular organisms.

Phylum Chlamydobacteria

Chlamydobacteria are the mycelial bacteria. The bacterial cells of a colony are arranged end-to-end. Their secreted extracellular sheath material forms thin mycelial filaments. This group of bacteria are an important source of three antibiotics, streptomycin (from Streptomyces), aureomycin (from Aureomyces), and actinomycin (from Actinomyces).

Phylum Spirochaetae

Spirochaetes are spiral rod-shaped bacteria that move in an undulating fashion. Many of these species are obligate parasites, that is, they cannot survive outside of the host cell. The most infamous spirochaete is *Treponema pallidum*, the syphilis-causing bacteria.

Phylum Mycoplasmae

The **mycoplasma** are among the smallest living organisms. Smaller than some viruses, they are intracellular animal parasites. They lack a cell wall and are therefore resistant to penicillin.

Phylum Cyanobacteria

The **cyanobacteria** have, until recently, been termed *blue-green algae*. Twenty-five hundred species of cyanobacteria have been classified to date. They are thought to be the oldest form of life on earth. The term blue-green algae is misleading; they also come in black, brown, purple, and red. They live as individual cells, in colonies, even as multicellular organisms.

The cyanobacteria are aerobic photosynthetic autotrophs. Their primary photosynthetic pigments are chlorophyll, and alpha and beta carotene. These organisms exhibit a flexibility unlike other photosynthetic organisms. They manufacture differing amounts of photosynthetic pigments depending on the available wavelength of light. This *chromatic adaptation* accounts for color variation within a species.

CHAPTER REVIEW

1. List the major differences between prokaryotic and eukaryotic cell structures.

2. Classify prokaryotes according to cell shape.

3. Differentiate between Moneran chemoheterotrophs, chemoautotrophs, and photoheterotrophs.

4. Define the major Moneran subdivisions, include characteristics and examples.

---

[1] See Chapter 8 for review of conjugation and binary fission in prokaryotic cells.

[2] Microscopic measurements are described in Appendix A.

[3] The exact classification schema varies with individual bias.

KINGDOM PROTISTA

Present systematic classification divides the protists into two groups: the photosynthetic protista and the protozoa.

PHOTOSYNTHETIC PROTISTA

The photosynthetic protists are unicellular, photosynthetic eukaryotes. Among these protists are the **phytoplankton**, called the "grass of the sea". Phytoplankton produce more new organic material per year than all other photosynthetic organisms combined! The photosynthetic protists are subdivided into three groups.

Phylum Pyrrophyta

The **Pyrrophyta**, or **dinoflagellates**, show considerable variety in cell size, shape, and metabolic activities. Ninety-three percent of the species are free-living, photosynthetic marine organisms. A few are parasites or symbionts. The characteristic whirling motion of the dinoflagellate is caused by two flagellar "whips." These whips are housed in grooves, one spanning the length of the cell and the other girding the midsection. Except for the grooves, dinoflagellates are covered by overlapping chitin plates. A number of dinoflagellates are *bioluminescent*, in which the enzyme luciferase changes ATP chemical energy into light energy.

The dinoflagellates exhibit many cellular peculiarities. The DNA chromosome remains permanently condensed and attached to the nuclear membrane. The nuclear membrane is a single membrane, in contrast to most eukaryotic double membranes, and it does not break down during mitosis. Meiosis in dinoflagellates, unlike other eukaryotic meiotic events, involves one, not two divisions. These unique features indicate early evolutionary divergence from other eukaryotic forms.

Phylum Euglenophyta

The **Euglenophyta** get their name (*eu* = true; *glene* = eye) from the striking presence of a red pigment shield or "eyespot" adjacent to a light sensitive area. When the eyespot is not covered, a single long flagellum propels the organism forward in search of light for use in the photosynthetic process.

Euglenoids are asexual. They lack rigid cell walls; however, the **pellicle**, a thin envelope outside the plasma membrane, provides them with some shape. Some euglenoids ingest solid food particles with an anterior **gullet**, a mouthlike depression. All species possess the gullet structure, but no evidence indicates that all ingest solid food. Euglenoids store their food as a unique carbohydrate, *paramylon*, found no where else among the photosynthetic organisms. They live in freshwater, especially in sources polluted with organic waste.

Phylum Chrysophyta

The **Chrysophyta**, golden plants, get their name from high cellular concentrations of carotene. This diverse group includes the yellow-green algae, the golden-brown algae, and the diatoms. All species are photosynthetic, although some amoeboid species also ingest solid food. Most species are aquatic; however, some yellow-green algae are found growing on soil, rocks, and trees, blown there by the wind. A few genera live in filamentous colonies; most are solitary.

Diatoms are the most abundant marine photosynthetic organism. Diatoms exhibit radial, biradial, or triradial symmetry. They are nonflagellated cells living inside a glass house made of pectin and silicon. Their glass shell resembles the Petri dish, the inner half fitting inside the outer half. They are intricately perforated with tiny holes allowing gas and water exchange. Deposits of diatom shells, *diatomaceous earth*, are used as abrasives in insulation, poool filters, toothpaste, and polishing materials.

Diatoms reproduce both sexually and asexually. During asexual reproduction, binary fission of nuclear and cytoplasmic elements takes place. Each new daughter cell receives one of the silicon halves from the parent cell. This becomes the outer lid and a new inner half-box is secreted. The daughter cell receiving the inner half lid from the parent will be smaller than the other daughter cell. Eventually a critically small diatom undergoes meiosis. Three of the four haploid nuclei degenerate. The remaining haploid nucleus grows flagella and swims free of its glass shell. Fusion with another flagellated spore forms a new diploid **auxospore**, (auxo = increasing) so named because it secretes a new, full-sized glass house.

PROTOZOA

The **Protozoa** are the animallike protista. The four major phyla are based on the manner of locomotion or lack thereof.

Phylum Mastigophora

The **Mastigophora** are the *flagellated protozoans*. The body structure of most mastigophorans reveals definite anterior (head) and posterior (tail) ends. The cell membrane is the only body covering for most species. The number of flagella varies from species to species. Nutrients are obtained in several ways; some species absorb nutrients through the plasma membrane and outer covering; others seek and capture prey, ingesting them by phagocytosis. Flagellates are not presently known to undergo sexual reproduction. Simple binary fission of nuclear material and longitudinal division of cytoplasmic elements takes place during asexual reproduction.

Many mastigophoran species are parasitic. Most notorious is *Trypanosoma gambiense*, the flagellate responsible for African sleeping sickness. Another species, *Trichonympha*, forms a mutualistic symbiosis within the intestines of termites. There it digests cellulose, passing nutrients back to the termite.

Phylum Sarcodina

Members of the **sarcodine phylum** exhibit amoeboid movement, flowing of cytoplasm into temporary extensions of the body, called **pseudopodia**. Some aquatic species carry intracellular endoplasmic oil droplets used for maintaining buoyancy while they float. Sarcodines typically feed through phagocytosis. Pseudopodia function to capture, surround, and engulf the food.

There are four groups of sarcodines, classified according to body shape. The **amoebas** exhibit an amorphous body shape characterized by numerous rounded pseudopodia. Some possess shells (tests) of silicon, chitin, and even bits and pieces of minerals glued together. The **foraminiferans** are marine organisms with interconnected, threadlike pseudopodia. A calcium carbonate shell of one or many chambers surrounds the cell body. The **heliozoans** are primarily fresh water organisms. The needlelike pseudopodia radiate from the cell body. They function not in locomotion but in capturing prey. Many species of heliozoans have external cell coverings composed of secreted silicon or foreign matter embedded in a gelatinous covering. All **radiolarian** species are planktonic marine organisms. Their body structure is spherical. Perforations in the chitinous capsule membrane allows cytoplasm to stream out. Radiolarians digest food outside of their body shell and the nutrient monomers pass through to the cell.

Some sarcodines reproduce only through asexual binary fission. Flagellated gametes are produced in those species that undergo meiosis and sexual reproduction.

Phylum Sporozoa

The **sporozoans**, with sporelike infective stages, are parasitic protozoans. Their life cycle is complex and usually involves sexual and asexual generations that develop in different hosts. The gametes of some species are flagellated. *Plasmodium*, the infectious agent of malaria, belongs to this group.

Phylum Ciliophora

The **ciliophorans** are the largest and most homogeneous group of protozoans. All possess **cilia** for locomotion or food acquisition. Cilia are more numerous and shorter than flagella; however their molecular substructure appears to be the same. A major characteristic of the ciliates is the presence of two types of nuclei, a vegetative (non-reproductive) **macronucleus,** and a reproductive **micronucleus**. Micronuclei vary in number from one to eighty. They contain the DNA material that functions in genetic exchange and reproduction. They give rise to the polyploid macronucleus, which specializes in RNA transcription and control of cellular differentiation. The majority of ciliates are free-swimming solitary organisms, although colonial and sessile species are also known. Outside of the plasma membrane is a complex covering called the **pellicle**. Embedded at right angles to the body within the pellicle of some ciliate species are explosive **trichocysts**. The trichocyst consists of a long shaft and an attached barb. It is used for anchoring the protozoan when it feeds, for defense, or for capturing prey.

Free-living ciliates engulf solid food particles. A mouth, or **cytosome**, opens into the **cytopharynx** canal. Food vacuoles are formed at its terminus, detach, then move throughout the cytoplasm where digestion takes place.

Asexual reproduction in ciliates is accomplished by transverse binary fission. In some species, budding occurs rather than division by cytoplasmic constriction. Conjugation, fusion of cytoplasm, reorganization and exchange of genetic material in the micronuclei, and meiosis occur during sexual reproduction.

CHAPTER REVIEW

1. List the two major divisions of the protista kingdom.
2. List three groups of photosynthetic protista and describe an organism belonging to each group.
3. Describe the protozoan groups in terms of locomotion.

KINGDOM FUNGI

The **fungi** are nonphotosynthetic, multicellular eukaryotic microorganisms. They have rigid cell walls. They feed by absorption of organic compounds: **saprophytes** feed on dead materials; **parasites** feed on living organisms. Familiar fungi include mushrooms, bread molds, *Penicillium* mold, mildews, and plant rusts.

CELL STRUCTURE AND METABOLISM

The cytoplasm of fungi is enclosed in branching tubes, or **hyphae**, and moves freely from cell to cell. This multinucleated mass is called **coenocytic**. Some fungi are **septate** (possess individual tubular cells); however, pores in the crosswalls (septae) allow cytoplasm and cell organelles to flow through. Others are **nonseptate** (without cell crosswalls). Cell wall composition varies from species to species. Common constituents are chitin, cellulose, and other polysaccharides.

Hyphae of many fungal cells form a filamentous mat, the **mycelium**. Fungal growth occurs by extension of the hyphal tubes. The growing mycelium mat spreads over and penetrates the substrate surface (the surface on which the fungus is living). Mycelium functions as a feeding structure. Digestive enzymes are secreted out of the cell into the substrate, and the monomer food units are then absorbed through the cell wall.

Yeasts and some aquatic fungi are unicellular. Other pathogenic fungi are dimorphic, exhibiting both unicellular and multicellular forms.

REPRODUCTION

The methods of reproduction differ for the unicellular and multicellular fungi. However, sexual and asexual modes of reproduction are common in both. Asexual reproduction increases the number of individuals within the species, whereas sexual reproduction permits genetic exchange, increasing the genetic diversity of the species. Sexual reproduction occurs during times of environmental stress.

Reproduction in Yeast

Saccharomyces cervisiae, a typical yeast, undergoes asexual reproduction during the budding process. Doubling of the DNA occurs while a cytoplasmic extension (bud) appears on the parent cell. The newly synthesized nuclear material moves into the bud. Nuclear separation is followed by cell division.

Sexual reproduction occurs when two haploid cells of opposite mating types fuse. See Figure 15.1 for the life cycle of a typical yeast.

Reproduction in Multicellular Organisms

Most multicellular fungi produce spores for dispersal or for protection of the species during environmental extremes. Asexual spores are dessication and radiation resistant. However, they are not heat resistant and show no dormancy. Sexual spores are usually heat resistant, often exhibit dormancy, and show signs of germination only after an activation process.[1]

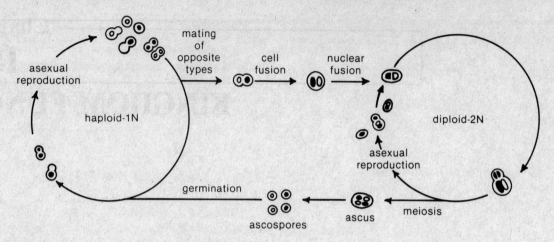

Figure 15.1. Sexual and asexual reproduction in *Saccharomyces cervisiae*.

Asexual Spores. Asexual spores are formed in one of three ways (Figure 15.2). **Chlamydospores** develop as specialized cells within the hypha. They form a thick wall, round up and detach from the hypha. **Conidiospores** are specialized cells produced on distinct hyphal extensions, **conidiophores**. They often occur in chains and clusters. **Sporangiospores**, formed within a special spore sac (the **sporangium**) are released when the spore sac ruptures.

HABITATS AND RELATIONSHIPS WITH OTHER ORGANISMS

Fungi exhibit a wide range of habitats. Freshwater, marine, and terrestrial species play a variety of roles.

Decomposers

The breakdown of dead organic material is known as **decomposition**. Their saprophytic way of life classifies fungi among the decomposers. In soil, the monerans, protists, and fungi play a vital role in **mineralization** — the decomposition of plant and animal remains and the return of the organic carbon back to its inorganic state. Without the mineralization process, all carbon would remain in the organic state, unavailable for recycling through organisms.

Symbionts

Some species of fungi enter into *symbiotic* relationships with other organisms. **A mutualistic** relationship, one which benefits both organisms, exists in **mycorrhizae**. The hyphae of soil fungus penetrate the root

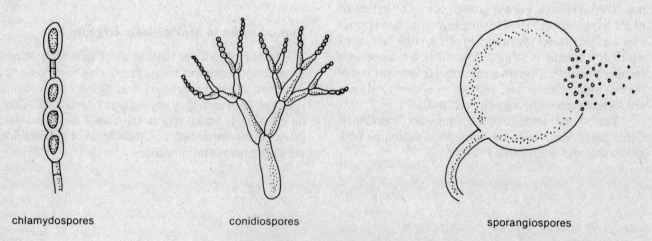

chlamydospores   conidiospores   sporangiospores

Figure 15.2. Asexual fungal spores.

hairs of plants, providing the root with greater surface area for absorption. The plant provides organic nutrients to the fungus. This relationship is thought to be almost universal. Plant growth that occurs without soil fungi is very poor. The most famous mycorrhiza is the truffle, a prized food of the gourmet.

Lichens are mutualistic relationships between fungi and algae. (There are estimated to be 15,000 different species of lichens.) Green algae cells are dispersed among the fungal hyphae. The photosynthetic algal cells provide nutrients while the fungus provides inorganic nutrients and dessication resistance. Lichens can grow on surfaces that do not permit growth of other organisms. Acids released by lichens that grow on rocks cause chemical weathering and over long periods of time contribute to the formation of new soil.

Industrial Uses of Fungi

Many manufacturing processes rely on the metabolic activities of fungi. Carbon dioxide and alcohol, the end-products of fermentation in yeasts (*Saccharomyces*), are used in the baking, wine, and beer industries. Penicillin is secreted by the *Penicillium* mold. In nature, penicillin protects the fungi from bacterial invaders. This antibacterial activity offers the same protection to humans. Other species of blue-green mold add the colors, flavors, and textures to Roquefort, Camembert, Gorgonzola, and Stilton cheeses. A metabolic end-product of the *Aspergillus* mold is citric acid. Many lemon-flavored candies, drinks, and medicines get their sour taste not from lemons but from the *Aspergillus* mold. This mold is also used in making soy sauce and saki.

Plant Parasites

Fungi are among the most destructive plant diseases. Because plants are a food source for both man and animals, fungal diseases have widespread effects. The disastrous Irish potato famine of the 1840s, which left 250,000 people dead, was caused by the fungus *Phytophthora infestans*.

Parasitic plant fungi are of three types. The short hyphae of **powdery mildews** extend into plant leaves to absorb nutrients. **Smuts** attack cereal grains. They reproduce rapidly and can be identified by their dark, dusty, smelly spores. Smut spores enter seedlings and the growing hyphae invade the plant stem. Parasitic **rusts** can affect large areas in short periods of time. Their life cycle involves growth on two different host organisms. These fungi are easily identified by the rust color of their spores on the host.

Human Parasites

Dermatophyte fungi grow on the skin and scalp, causing ailments such as dermatitis, ringworm, athlete's foot, and jock itch. Soil spores cause Valley Fever (coccidiomycosis) and histoplasmosis, two lung diseases commonly confused with tuberculosis on Xrays.

TAXONOMY

Within the diversity of the fungi, exceptions exist to every rule or classification schema. The three subkingdoms within the Fungi Kingdom will be discussed. These have been classified on the basis of cellular structure and composition, and growth and reproduction characteristics.

Subkingdom Gymnomycota

The **Gymnomycota**, or **slime molds**, are further subdivided into two groups, the acellular and the cellular slime molds.

Acellular Slime Molds. The multinucleate plasmodium of the **acellular slime mold** looks like a shiny amoeba. In a dry habitat, its amoeboid movements cease and haploid spore-bearing sporangiophores develop. Air currents carry the spores to new locations. Under favorable conditions, the spores develop into motile haploid cells that fuse to form the diploid cell. Mitosis occurs and the cytoplasmic mass increases without cell division, producing another plasmodium.

Cellular Slime Molds. Cellular slime molds do not go through the plasmodium stage of development. Free-living independent cells divide mitotically and proliferate when environmental conditions are favorable. Under adverse conditions, certain amoeboid cells secrete *cyclic AMP (cAMP)*, which acts as a chemical messenger, causing cell aggregation and the transformations seen in Figure 15.3.

Subkingdom Dimastigomycota

The **Dimastigomycota**, or **water molds**, produce nonseptate mycelium of only a few cells. The aquatic "parasites" only invade living organisms already traumatized by disease or injury. Terrestrial species include the powdery mildews and blights (the species responsible for the Irish potato famine, for example). Hyphae of the blight fungi grow through the pores on moist leaves. The mycelium spreads through the spongy

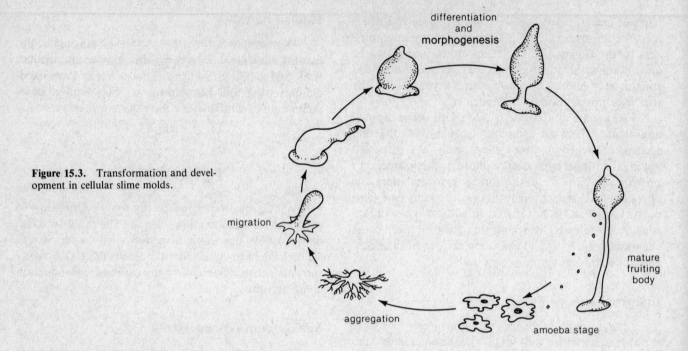

Figure 15.3. Transformation and development in cellular slime molds.

layer of the leaf, penetrates photosynthetic cells and absorbs the nutrients as they are manufactured.

Flagellated spores develop within asexual sporangia. New colonies develop when released spores encounter nutrient sources. Sexual reproduction is diagrammed in Figure 15.4. Growth of specialized thick hyphae results in the formation of **oogonia** – spherical structures which produce egg cells. **Antheridia,** specialized sperm producing hyphae, grow over the oogonia and send out fingerlike fertilization tubes.

Subkingdom Eumycota

The **Eumycota** or **"true fungi"** are composed of four divisions (phyla).

Phylum Chytridiomycetes. The **chytrids** are a diverse group in their appearance. Some species are unicellular without an extensive mycelial mat; others possess hyphae and flagellated spores. Terrestrial forms are plant or fungal parasites. Some aquatic species are algal parasites.

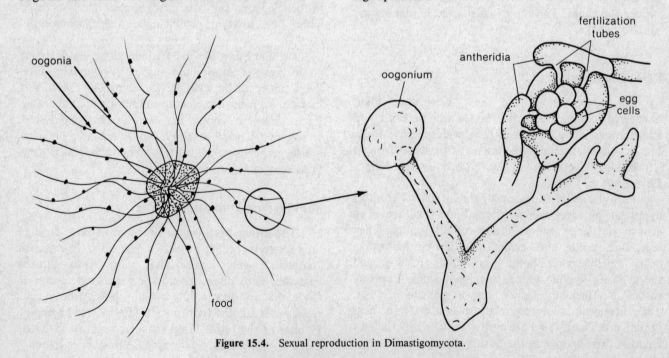

Figure 15.4. Sexual reproduction in Dimastigomycota.

Phylum Zygomycetes. The **Zygomycetes** are non-septate fungi. The hyphae within the mycelial mass lack regular form. Most species within this division are saprophytes. The common bread mold, *Rhizopus*, is a zygomycete. The tiny black dots that appear on a moldy mass are the asexual sporangia. They grow on the end of specialized vertical hyphae, the sporangiophores. This arrangement contributes to the mold's typical fuzzy appearance.

Sexual reproduction, shown in Figure 15.5, occurs beneath the substrate on which the mold is growing. No specialized male or female structures are formed. Com-patible mating types are designated plus (+) and minus (−). Mycelial nuclei are haploid in genetic content. Fusion of haploid nuclei within the **gametangia** produce the diploid zygospore. The diploid stage is short lived. Meiosis reduces the nuclear content to the haploid state. All but one haploid nuclei within the gametangia degenerate. This cell germinates and produces a new mycelial mass.

Phylum Ascomycetes. The **Ascomycetes**, or **sac fungi**, are septate fungi (Figure 15.6). Long chains of asexual conidiospores are produced at tips of specialized

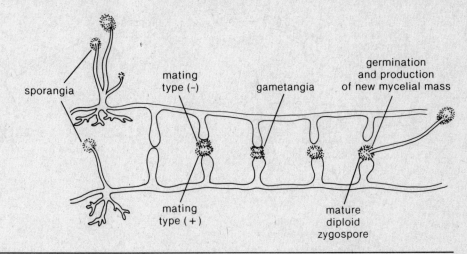

Figure 15.5. Sexual reproduction in Zygomycetes.

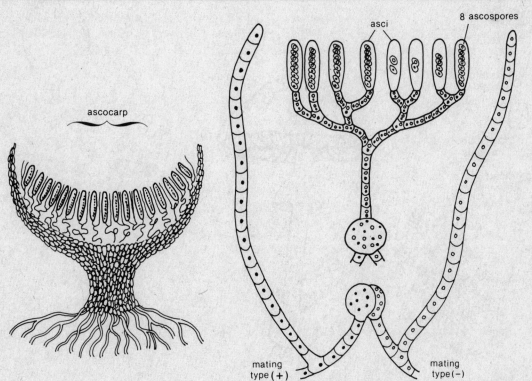

Figure 15.6. Sexual reproduction in the sac fungi.

hyphae, the conidiophores. Sexual reproduction is more diverse, however. The name ascomycetes is derived from the sac-like structure, the **ascus**, which houses the sexual **ascospores**. A group of asci form a fruiting structure, the **ascocarp**. Fusion of the multinucleate hyphal swellings of opposite mating types (+ and −) results in the formation of a **dicaryon**. Fusion of the nuclei does not occur immediately. Many binucleate hyphae emerge, each containing one + and one − nucleus. Ascocarp development, characterized by the formation of a sterile hyphae covering, signals asci maturation and fusion of the nuclei. Meiosis proceeds creating four

haploid nuclei. One sequence of mitosis increases the number of of sexual ascospores to eight per ascus.

Ascomycetes include blue and green molds of citrus fruit, powdery mildews, and yeasts.

Phylum Basidiomycetes. The **Basidiomycetes**, or **club fungi**, are septate fungi. They include the parasitic rust and smut species, as well as the more familiar mushrooms and toadstools. In those species that parasitize two different hosts, each host serves a different role in the sexual/asexual reproduction cycle. Figure 15.7 outlines the sexual and asexual aspects of

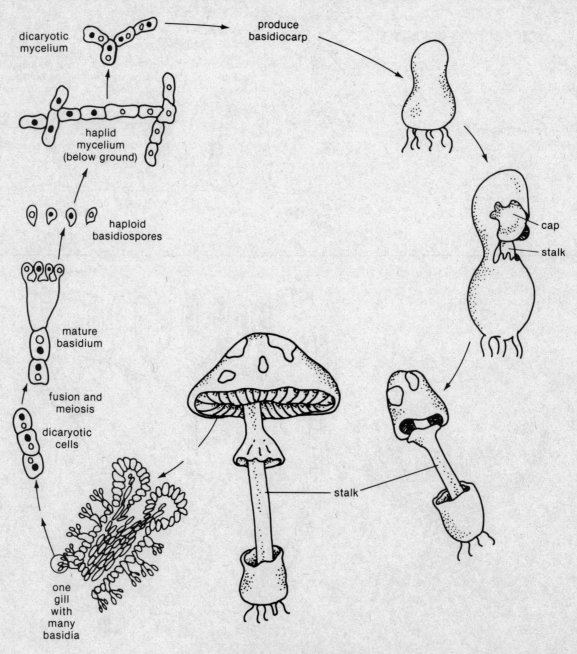

Figure 15.7. Life cycle in the mushroom.

the life cycle of a typical basidiomycete, the mushroom. The fungal mycelial mass is never seen above ground. The structure we call the mushroom is the reproductive sporeproducing structure, the **basidiocarp**.

Phylum Deuteromycetes. Members of the **Deuteromycetes** are the **"imperfect fungi."** The name of this group of fungi refers to the lack of sexual reproduction within the approximately 24,000 known species. Little taxonomic work has been done on this group of fungi. A number of human parasites belong to this group.

CHAPTER REVIEW

1. Describe the body structure of unicellular and multicellular fungi.

2. What is the difference between sexual and asexual reproduction in the fungi? What role does each play in the life cycle?

3. What are spores? What function do they serve?

4. What role do the fungi play in decomposition? Why is it important?

5. Give an example of
   a. a fungal mutualistic symbiont
   b. fungi used as a food source
   c. a fungal plant parasite
   d. a fungal human parasite

6. Describe the life cycle of
   a. cellular slime molds
   b. a common zygomycete
   c. a common ascomycete
   d. a common basidiomycete

[1]Sexual reproductive structures and spore formation are distinct for each taxonomic group of fungi. They are discussed in more detail later in the chapter.

KINGDOM PLANTAE

The members of the **Kingdom Plantae** are multi-cellular photosynthetic organisms. They are sexual reproducers (primarily) with alternating haploid (1n) and diploid (2n) generations. A few single-celled photosynthetic organisms are classified as plants. Many exceptions, however, are the result of secondary loss or modification as isolated groups continued to adapt to their changing environment.

TAXONOMY

Plants can be divided into three subkingdoms based on photosynthetic pigments, structure, and carbohydrate storage products. (In botanical nomenclature, the term "division" is used rather than "phylum.")

Subkingdom Rhodophycophyta

Most members of Subkingdom Rhodophycophyta are marine organisms. The plants within this subkingdom are called the **red algae** when, in fact, they are often green, black, blue or violet. Similar to cyanobacteria, many species exhibit **chromatic variation** — different amounts of pigments are produced in individuals at different depths, resulting in color variation within a species. They utilize chlorophylls a and d, phycocyanin ("algal blue-green") and phycoerythrin ("algal red").

Red algae grow attached to the substratum by a holdfast, a type of anchor, not a true root. Food is stored principally in the form of *floridean starch*. Another polysaccharide storage product, *agar-agar*, is utilized in bacterial growth media, in the medical and dental fields, in cosmetics, and as a food additive.

Reproduction in the red algae exhibits definite **alternation of generations**, the haploid sexual generation follows a diploid asexual generation. The diploid sporophyte plant undergoes meiosis and produces spores. The spores germinate and grow into multicellular haploid gametophytes. The male and female organisms produce haploid sex cells (gametes). Fertilization produces a new diploid sporophyte. The gametophyte is considered to be the sexual generation; during this generation sex cells are produced, albeit by mitosis, not meiosis.

All organisms within this subkingdom are grouped into a single division, **Division Rhodophyta**.

Subkingdom Phaephycophyta

The marine **brown algae** are characterized by **fucoxanthin**, a brown pigment. (All brown algae are actually brown.) They utilize chlorophylls a and c in the photosynthetic process. Reproductive sperm and egg cells are flagellated. Brown algae range in size from microscopic filaments to the giant kelps, 30 meters tall. The latter are the most specialized brown algae; their tissues and organs include a **holdfast** (anchor), **stipes** (stemlike structure), **blades** (resembling flat leaves), and **bladders** (hollow spherical floats). Carbohydrate storage forms are *laminarin* and *mannitol*. A characteristic cell wall structural polysaccharide is *algin*, used commercially in ice cream.

Reproduction in some species of brown algae is characterized by the absence of the multicellular haploid stage. This form of reproduction is rare in plants, but similar to that found in the Kingdom Animalia. In other brown algae, the diploid sporophyte generation is the dominant plant. The haploid gametophyte may be microscopic, only a few cells, whereas the sporophyte reaches sizes of several meters.

All organisms within this subkingdom are grouped into a single division, **Division Phaeophyta.**

Subkingdom Euchlorophyta

The organisms within this subkingdom are more diverse in size, body structure, and location. However, all euchlorophytes use chlorophylls a and b and carotenoid pigments in the photosynthetic process. They store carbohydrates as *amylose* and *amylopectin.*

Division Charophyta. The **stonewarts** of today retain many structures found in their Silurian fossil ancestors. Differentiating them from other euchlorophytes are the calcium salts incorporated into the cell walls. These marine plants are bottom dwellers, found at depths up to 140 meters.

Division Chlorophyta. The chlorophtes are the **green algae.** This diverse group includes unicellular, colonial, and multicellular species; flagellated and nonflagellated species; body structures of chains, filaments, and flattened blades. Examples of chlorophytes include: **Chlamydomonas,** with its light sensitive red eyespot; the colonial **Volvox;** the coenocytic siphonous algae, **Acetabularia,** which grows as a single cell reaching five to nine centimeters in height; **Codium** (dead man's fingers); and the truly multicellular **Ulva** and **Ulothrix.**

Division Bryophyta. Living bryophytes include the **mosses, liverworts,** and **hornworts.** Along with all previously mentioned plants, they are **nonvascular plants;** that is they lack specialized transport and support systems. Ancestral bryophytes are thought to be the first life form to successfully make the transition from an aquatic to a terrestrial way of life.

All bryophyte species are multicellular and have specialized tissues. The **rhizoid** anchors the plant in the substratum while drawing in water and minerals. A **central cylinder** surrounded by **cortical cells** and the photosynthetic epidermis forms a simple structural support "stem." The main body of the plant is the **thallus,** composed of **leaf scales.** Although not as complex, the flat ribbon of photosynthetic cells functions as a "leaf." The outer epidermal cell layer in some species is coated with a moisture barrier, **cutin.**

Reproduction in the Bryophytes resembles reproduction in marine plants. Sperm are flagellated and require water to swim. All bryophytes show alternation of generations with a dominant haploid gametophyte. The fertilized gamete is protected from dessication by a shield of cells, the **archegonium.** Some mosses are

monoecious (produce egg and sperm in the same thallus); most are **dioecious** (separate sexes, only one type of gamete produced in a given thallus). Some liverwort species undergo asexual reproduction; **gemmae,** reproductive buds, dislodge and form a new plant.

Division Trachaeophyta. The trachaeophytes are **vascular plants** with conducting tissue. These tissues include **root systems, stems,** and **leaves.** Two reproductive modifications have evolved within the trachaeophytes: *nonflagellated sperm,* which do not require water for translocation to the egg, and *seeds.*

There are five major groups of trachaeophyte plants. The phylogeny and organization within this division are still under scrutiny. One classification schema is presented below:

Subdivision Psilophyta are the **naked plants.** They have no true roots but have a rhizoid system. They contain vascularized stems. Leaves are scalelike. Sperm are flagellated and require water for translocation. Psilophytes are abundant in the fossil record of the mid-Paleozoid era.

Subdivision Lycophyta represents the **club mosses.** They have true **adventitious roots,** emerging from the adult stem. The vascular system is well-developed. Their leaves are small and spiral around the stem. Some species produce **runner stems** that send out additional roots and vertical stems. The haploid sporophyte generation is dominant. Fertilization requires the presence of water. Lycophytes were the dominant plant of the Devonian and Carboniferous period and grew to be 50 meters in height, 2 meters in diameter. Decomposition of these fossil remains formed the coal, gas, and crude oil used today.

Subdivision Sphenophyta are the modern-day **horsetails.** Like their Lycophytan counterparts, they dominated the Carboniferous forests. These plants deposit silica in their epidermal cell walls. The dominant diploid sporophyte plant takes two forms. One is a vegetative shoot, the other a nonphotosynthetic spore producing shoot. The spores germinate into separate male and female gametophytes the size of a pinhead. Fertilization is water dependent.

Subdivision Pterophyta are the **ferns.** The underground **rhizome** of the dominant diploid sporophyte gives rise to fine **roots.** Single leaves, divided into **leaflets** emerge from the ground. Spores are produced in **sori,** reproductive structures on the underside of the leaflets. Haploid spores germinate into the heart-shaped gametophyte **prothallium.** Its sperm-producing **antheridia** and the egg-producing **archegonium** are anchored by rhizoids. Water is required for fertilization. Many ferns have evolved a hormonal system that inhibits self-fertilization. Antheridial development is stimulated on a neighboring prothallium while their own similar development is inhibited.

Subdivision Spermophyta are today's **seed plants**. They have extremely well-developed water- and nutrient-drawing vascular systems. This enables many to live in desert environments. The dessication-resistant **male pollen cell**, and the seed, an embryo with stored food surrounded by a hardened seed coat, were new developments that enabled these species to dominate the terrestrial environment.

The two main groups within the Spermatophytes are **Gymnosperms** ("naked seed" plants) and the **Angiosperms** (the "flowering" plants). The Gymnosperms produce seeds without a fleshy surrounding fruit. These plants include the *ginkgo*, the *cycads*, and the *conifers*. The ginkgo and cycads produce flagellated, swimming sperm. A single conifer (cone-bearer) produces separate male and female cones. The leaves of conifers have been modified into needles. The outside layer of cells provide a hard, porous covering. Needles are shed at regular intervals, not all at once.

The Angiosperms have two structures that differentiate them from other plants, a flower and a fruit. (These structures are discussed later in this chapter.) The Angiosperms are classified as **Class Monocotyledonae (monocots)** and **Class Dicotyledonae (dicots)** based on the number of cotyledons (seed leaves). Monocots have a single cotyledon; the dicots have two. The monocots are the lilies, palms, grasses (cereal grains), and orchids. The dicots include herbs, flowering trees and shrubs, and many of the fruits and vegetables eaten by humans (Figure 16.1).

MECHANISMS OF TRANSPORT

Water intake occurs initially through osmosis. **Root hairs**, extensions of root epidermal cells, concentrate salts and minerals, creating an internal hypertonic environment. Water from the soil flows into the roots and then moves to the rest of the plant through xylem tissue. The **xylem system** is made of thick-walled xylem cells fused end to end, resulting in the formation of transport tubes. Water is moved throughout the plant by the free energy of evaporation or **transpiration** due to the cohesion-tension of the water molecules. The hydrogen atoms covalently bound to oxygen in the water molecule are electrostatically attracted to the oxygen of an adjacent water molecule. This forms a hydrogen bond. This secondary attraction (cohesion) produces tensile strength that can raise the water within a plant to its highest leaves. The concentration of water in the atmosphere is less than the concentration of water within the cells, and water molecules move from the leaf surface into the atmosphere (transpiration) via osmosis.

Gas exchange occurs in a variety of ways. Breathing spaces, **lenticels**, are found in the bark of woody plants. In addition, the leaves and stems have **stomata**, openings formed when the turgor pressure inside two **guard cells** is high. Gases and water can escape through these openings found mainly on the underside of leaves. Root hairs also function in gas exchange if the soil environment is not water saturated.

Minerals are actively transported across the plasma membrane of root hair cells. They are then carried up the plant by water flow within the xylem.

Nutrients are transported through the plant by the living cells of the **phloem system**. The phloem system is composed of living, thin-walled sieve tubes. Mature phloem cells lose their nuclei but retain their cytoplasm. They do not form a continuous tubelike transport system. Two separate theories have developed to explain the translocation process. The **mass flow theory** suggests a type of circulatory pattern within the plant between the phloem (containing concentrated sugars) and the xylem (carrying the water). Osmotic gradients force water out of the xylem and into the sugar-rich solution (sap) carried by the phloem. As the phloem vessels become more turgid (swollen by water pressure from within), the sap flows out of the leaf. As the fluid reaches other tissues, active transport moves the nutrients across the plasma membranes of the sieve-tube phloem cells and into the tissue. This creates a new osmotic gradient; water flows out of the phloem and back into the xylem and eventually returns to the leaves. Some experimental data supports this theory.

A second theory is the cytoplasmic streaming or cyclosis theory. Observation under the light microscope reveals tiny cytoplasmic currents within some plant cells. An actin and tubulin cytoskeleton is hypothesized to be responsible for this motion. It is theorized that nutrients are carried within these streams and are deposited at the end of one cell and then transported across the plasma membrane to the adjacent cell. However, most investigators have not observed cyclosis in mature phloem cells.

PLANT GROWTH AND DIFFERENTIATION

Unlike the higher animals, once a higher plant reaches maturity, it continues to grow. Embryonic meristematic tissue persists in the full-grown plant and is capable of developing into other tissues.

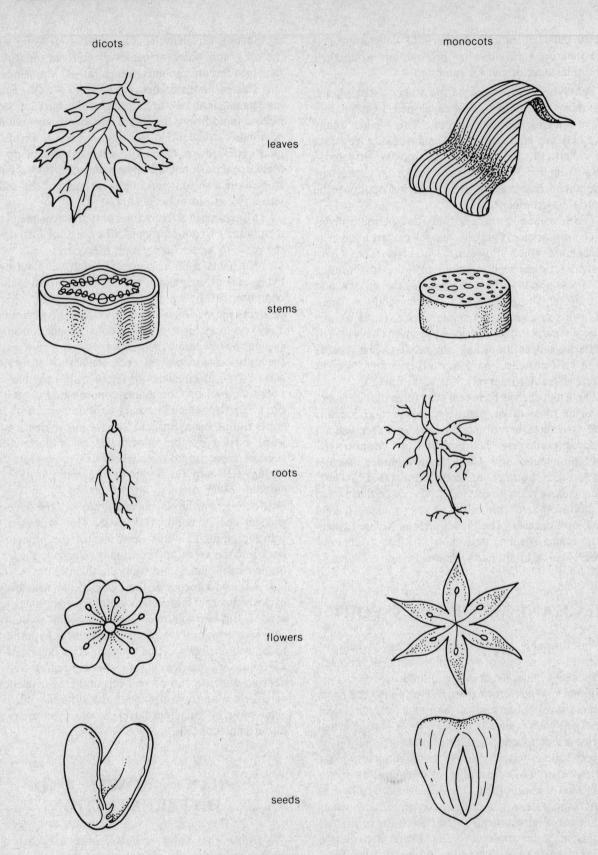

dicots   monocots

leaves

stems

roots

flowers

seeds

Figure 16.1.   Dicot and monocot plant structures.

Germination

The diploid seeds of many plant species undergo a period of quiescence or **dormancy**, remaining viable for many years. Requirements for seed **germination** (sprouting) vary from species to species. Some require temperature extremes (heat or cold); others are inhibited by chemicals produced from the decomposition of leaf litter; still others germinate only after passing through an animal's digestive system. Common germination requirements include oxygen availability, adequate water supply and favorable temperatures. Germination and early plant growth in both monocots and dicots is presented in Figure 16.2. The first steps in germination involve absorption of water, softening of seed coat, and activation of enzymes. The enzymes break down the food stored within the seed into monomers used for cell growth until the plant is capable of photosynthesis.

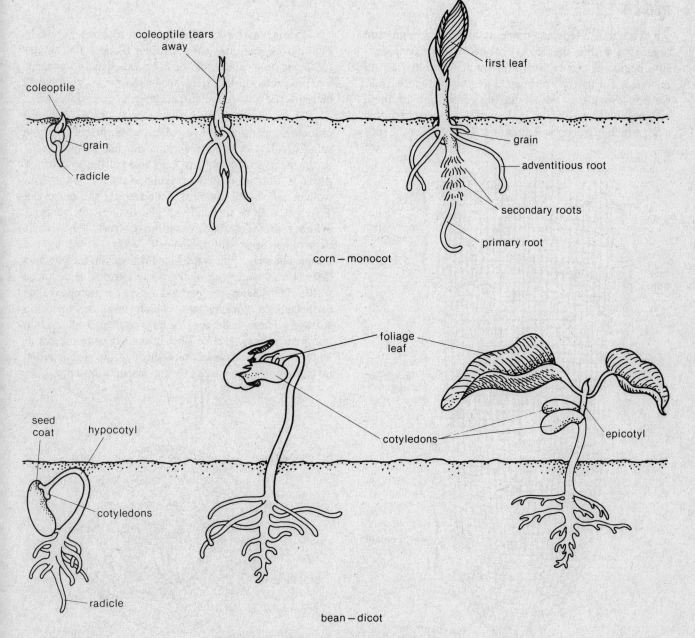

Figure 16.2.   Germination in monocot and dicot seeds.

The monocot seed remains underground, but the *cotyledons* of the dicot seed are lifted out of the soil at the end of the *hypocotyl*. The cotyledon and *coleoptile* provide similar functions, protection of the *epicotyl* leaves as they emerge through the soil. As the cotyledons open and the coleoptile unrolls, the epicotyl are exposed to the sun. Chloroplast development, cell division, and elongation proceed rapidly. By the time the endosperm food reserves of the seed are depleted, the embryonic leaves are photosynthesizing on their own. The cotyledons wither and drop.

Roots

The root meristem tissue is already present and organized within the seed. Post-germination cell division begins in this region. This undifferentiated *apical meristem* forms the *root cap*, pictured in Figure 16.3. Carbon dioxide given off during respiration of these cells combines with water in the soil to produce carbonic acid, which dissolves soil minerals. This enables the root

tip to grow forward. Cells of the root cap are worn away and replaced. Meristematic cell divisions push some daughter cells downward to form new root cap cells; others remain within the meristematic zone or are pushed upward where they undergo differentiation. Differentiation also occurs within the elongation zone. As cell divisions cease, differentiaion of sieve cells within the phloem system begins. Farther up the root the xylem vessel system is also beginning to differentiate. Differentiation of endodermal cells forms the cylinder around the xylem and phloem systems within the root. This differentiation occurs within the maturation region of the root.

Primary Growth. The mature dicot root, shown in Figure 16.4, contains three distinct tissues. The single-cell layer outer epidermis serves protective and absorptive functions. Long extensions of these cells, root hairs, increase the absorptive surface area of the roots.

The inner vascular tissues are the xylem and phloem, separated by **vascular cambium** cells. These small dividing cells differentiate to produce xylem cells to the interior and phloem cells toward the exterior of the root. Surrounding the xylem and phloem is the thin-walled, single-cell layer of endodermis, the **pericycle**. Together with its internal vascular tissues, this **vascular cylinder** provides rigidity within the root. These cells deposit *suberin* in their cell walls, forming the **Casparian strip**. This semiclosed waxy cylinder prevents diffusion of water out of the water permeable cellulose walls. The **cortex**, or ground tissue, is composed of unspecialized **parenchyma**, which functions in food storage. These cells are loosely packed and contain many air spaces. Unlike flow in the opposite direction, water and minerals freely pass through the parenchymal layer from the root hairs to the vascular system.

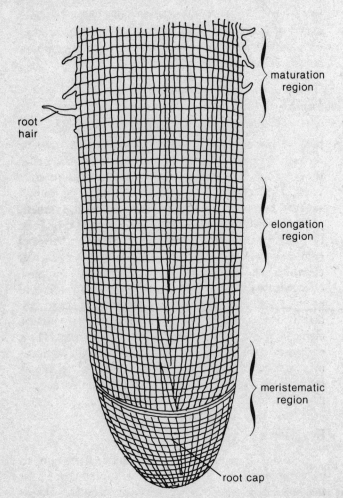

Figure 16.3. Longitudinal section of root.

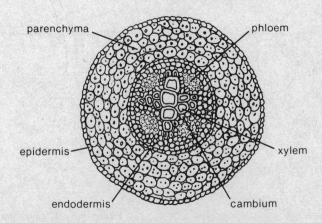

Figure 16.4. Cross section of mature dicot root.

Secondary Growth. Production of plant tissue by cell division of meristematic tissue other than apical meristem is called **secondary growth**; the tissues formed are **secondary tissues**. Dicots produce lateral meristem or vascular cambium within their roots; **secondary roots** develop from the pericycle tissue of such plants. In addition to new root structures, growth of the pericycle increases the thickness of the root.

Stems

The **stem** provides the transport route for nutrients from leaves to roots and for water and minerals in the opposite direction. It also provides support for the above-ground plant structures.

Primary growth. The **shoot** contains regions analogous to those found in the root: apical meristem, elongation zone, and maturation or differentiation zone. The parenchymal tissue within the stem is known as **pith**. The terminal bud contains the shoot apical meristem tissue that gives rise to new leaves and new branches — **leaf primordia** and **branch primordia**. Cell differentiation within the maturation zone produces tissues needed for structure and metabolism in the mature plant (Figure 16.5).

As the meristematic zone is pushed upward by the elongation of cells below it, leaf and branch primordia remain behind as **nodes**, regions of potential leaf and stem growth. **Lateral branch buds** are protected by modified leaves known as **bud scales**. Growth of lateral branch buds is inhibited by auxin produced by the apical meristem. As the apical meristem grows farther away, the concentration of auxin becomes less. The inhibitory effect is decreased and the lateral buds most distant from the apical meristem may sprout. This condition of growth is known as **apical dominance**.

Monocot grasses, which evolved under the selective pressure of animal grazing, do not develop in the same way. Loss of growing tips by grazing would forever stunt their growth. Instead, the dividing basal meristem tissue is found below the leaves and stem.

Secondary Growth. While monocot stems grow, they remain similar in structure to the one in Figure 16.5. Secondary growth in dicots, however, produces a stem structure similar to that in Figure 16.6. Dicot stems contain vascular cambium, the accessory meristematic tissue of secondary growth. As in dicot roots, the undifferentiated cells of the vascular cambium ring continue to divide. Cells on the ring's interior differentiate into xylem; those on the ring's exterior differentiate to form phloem. The phloem is arranged in triangular vascular bundles. The **rays** found within each year's annual xylem growth are vascular phloem elements. The **cork cambium** is a ring of meristematic tissues that produce cork cells. **Bark** is composed of the vascular cambium and all external tissues.

Leaves

The leaf contains extensions of the xylem and phloem tissues. Parallel **vascular bundles** enter the leaf through the **petiole**, the portion that attaches the leaf to the stem and branch out as leaf veins. Smaller vascular veins, enclosed by layers of parenchyma, form **bundle sheaths**. Surrounding large veins are nonphotosynthetic parenchyma cells. Photosynthetic mesophyll cells form two different layers within the leaf. Between the lower epidermis (the underside of the leaf) and the vascular elements is the **spongy layer**, loosely arranged mesophyll cells and numerous air spaces. The air spaces are connected to stomata. Sandwiched between the upper epidermis (the topside of the leaf) and the vascular elements is the **palisade layer** of mesophyll cells. These elongated cells form an orderly array that allows maximum cell exposure to the sunlight. **Cyclosis**, cycling of the palisade cell chloroplasts, further maximizes capturing of sunlight.

Regulation of Growth

Auxin. The plant hormone, **auxin**, is thought to be responsible for a plant's growth preferences, or **tropisms**. Examples are **positive or negative phototropism** — growth toward or away from light, and

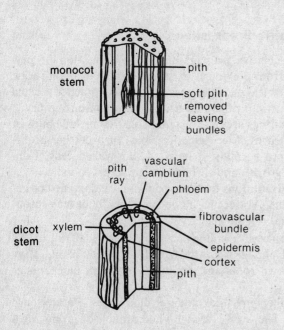

Figure 16.5. Sections of monocot and dicot stems.

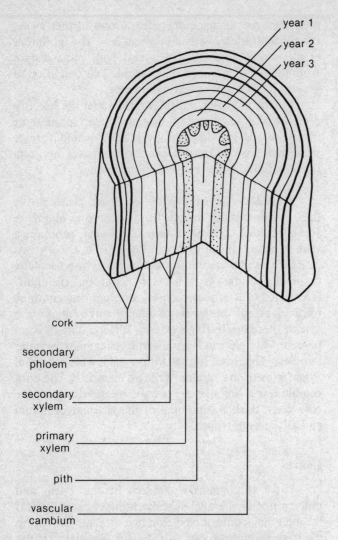

year 1
year 2
year 3

cork
secondary phloem
secondary xylem
primary xylem
pith
vascular cambium

Figure 16.6.   Cross section of mature dicot stem.

positive or negative geotropism — growth downward or upward. Concentrations of auxin causes cell lengthening. Laboratory studies have shown that auxin induces RNA synthesis. The steps linking RNA synthesis and cell elongation are not known.

Light causes the concentration of auxin to decrease. The mechanism is as yet unknown. Cells with low auxin concentrations do not elongate. In diffuse sunlight, the apical meristem continues to divide, but not grow in size. Farther down the shoot the concentration of auxin is greater and the cells elongate. In the case of positive phototropism, the cells of the side facing the sun have low concentrations of auxin and do not elongate. The cells of the side away from the sunlight have increased concentrations of auxin, elongate, and bend the growing tip toward the light source.

Gibberellins. Gibberellins are a family of molecules formed around the growing tip of young leaves. They play a role in stem elongation. "Dwarf" plants appear to have a decreased amount of chemically active gibberellin molecules. Gibberellins secreted by the germinating embryo also stimulate the synthesis of α-amylase. This enzyme breaks down starch to nutrient monomers within the seed.

Cytokinins. Cytokinins, in conjunction with auxins, stimulate cell division and differentiation. The ratio of these hormones determines what type of cellular differentiation takes place. Cytokinins are also associated with the process of **senescence** (aging) in plants.

Ethylene. Ethylene is a simple gaseous organic compound C_2H_4. It controls the ripening process and plays a role in a seedling's emergence from the soil.

Abscisic Acid. Abscisic acid functions in the control of leaf separation, **abscission**. Acting in conjunction with auxin, abscisic acid causes the cells at the base of the petiole, the abscission zone, to die and harden prior to leaf separation. When the leaf falls, the wound is already healed preventing excessive water loss. Abscisic acid is thought to induce winter dormancy in plants by inhibiting the actions of the growth hormones auxin and gibberellins.

ASEXUAL REPRODUCTION

Asexual reproduction produces new individuals genetically identical to the parent. This is a valuable horticultural and agricultural tool.

Vegetative Propagation

Vegetative propagation is the formation of new plants from previously existing portions of plant roots, stems, or leaves. Some species routinely practice asexual reproduction; others do it only in response to injury.

Cuttings are bits of stem tissue that include buds or entire leaves. Meristematic tissues of a cutting proliferate into a **callus**, a mass of thin-walled cells. Continued growth and differentiation forms roots.

Adventitious buds are common on damaged or cut off stems. Also found hidden on roots, buds may sprout and send out **suckers**.

Runners or **stolons** are thin stems that fall to the ground and grow away from the parent plant. They send down roots and eventually form an independent plant.

Tubers are thick underground stems. Potatoes are tubers. The "eye" is a lateral bud capable of producing a stem and adventitious roots.

Rhizomes are another kind of underground stem. They produce periodic nodes that send roots downward and a stem upward.

Grafting is done with the stems of woody plants. A callus develops around the vascular cambium of both the recipient and the inserted stem. The callus further differentiates into new xylem, phloem and vascular ambium.

Apomixis

Some species produce seeds without the union of gametes. This process is called **apomixis**. These seeds are genetically identical clones of the parent. Pollination in some of these plants acts merely to trigger the development of the endosperm seed food source.

SEXUAL REPRODUCTION

Sexual reproduction allows genetic variability in plants. The pool of genetic information may be activated by mutation at some future generation, providing an adaptation in a changed environment.

Flower Structure

Sexual reproduction in the Angiosperms takes place within the flower structure, as shown in Figure 16.7. All parts of the flower are modified leaves. The *sepals* and *petals* are accessory parts, not directly involved in reproduction. The male reproductive structure, the **stamen**, produces pollen grains within the *anther*. The female reproductive structure is the **pistil**, consisting of *ovary* and *ovule* cavity, *style*, and *stigma*.

Initiation of Flowering. The production of the flower structure occurs during a set developmental period. It does not appear to be a hormone initiated event; rather, it is dependent upon light and dark cycles of photoperiodicity. Laboratory experiments have shown that the amount of darkness is critical. A single brief exposure to light in the middle of the night can cause a long-day plant to bloom. **Long-day plants** flower before the summer solstice (June 21). They require a minimal amount of darkness to trigger the flowering process. **Short-day plants** require a greater amount of darkness and flower after the summer solstice. **Day-neutral plants** respond to a stimulus other than light or dark; some flower continuously.

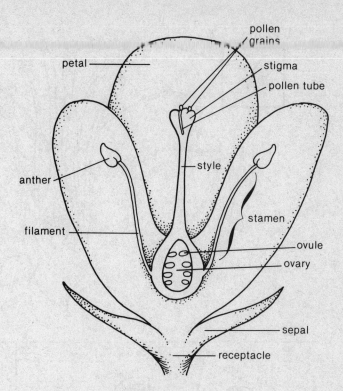

Figure 16.7. Flower structure.

Gamete Formation

Megasporgenesis is the formation of the egg cell along with its accessory embryo sac is diagrammed in Figure 16.8. Within the ovule of each pistil is a large diploid **megaspore mother cell**. Meiosis occurs, producing four haploid cells. Three disintegrate and the single haploid megaspore cell undergoes three successive mitotic divisions, resulting in the formation of a **megagametophyte** with eight haploid nuclei. Four nuclei migrate to each end of the cell. One from each end then migrates back to the center. Cell plates form, the cytoplasm is divided, and seven cells unequal in size are formed. The large cell in the center contains the two haploid nuclei. This is the endosperm mother cell. It will form seed endosperm after fertilization. The seven cell structure is called the **embryo sac**. A single haploid cell located at the exposed end of the ovule is the egg cell. The egg cell and the endosperm mother cell are both fertilized by pollen cells, a process known as **double fertilization**.

A similar set of events occurs within the anther during **microsporogenesis**, the formation of pollen cells (Figure 16.8). Numerous diploid microspore mother cells undergo meiosis within four pollen sacs. The haploid microspores produced replicate and divide during mitosis. A **generative nucleus** and a **tube nucleus** are formed. A tough coat is formed around each set of nuclei. This is the pollen grain structure.

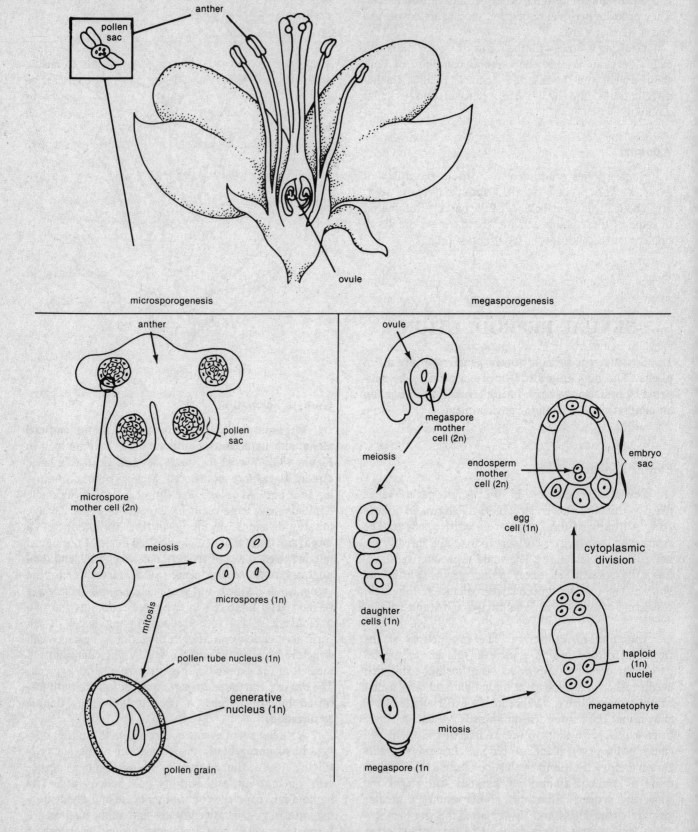

Figure 16.8. Details of gamete formation.

Pollination and Fertilization

Plants have evolved a variety of means to ensure that **pollination** takes place. During this process, pollen grains are transported from the anther to the stigma. In self-pollinators, pollen from the anther is transferred to the stigma of the same plant. More common is **cross-pollination**; pollen from one flower ends up on another flower. Both methods of pollination utilize a variety of external agents — the wind, insects, birds, even bats, accomplish pollen transport. Animals inadvertently pollinate flowers. Lured by nectar or by the pollen, an animal may remove pollen grains from the anther and relocate the grains to a stigma itself or to another animal. Flowers have evolved a variety of ways to attracts pollinators — colors, odors, even flower shapes may lure.

Sticky substances coating the flower stigma trap pollen grains and stimulate them to germinate. A long thin-walled **pollen tube** secretes enzymes and grows down into the ovary. The haploid tube nucleus moves along with the growing tip of the pollen tube. The haploid generative nucleus undergoes mitosis and produces two sperm nuclei. These nuclei participate in double fertilization; one fuses with the egg cell and forms the fertilized egg or zygote; the other fuses with the two haploid nuclei of the mother endosperm cell to form the triploid (3n) endosperm.

Development of Fruit

After fertilization, those parts of the flower not needed for cell maintenance degenerate. The ovary swells, forming the the structure called the **fruit**. Enclosed within are the fertilized eggs, the **seeds**.

Seeds. A variety of mitotic activity occurs after fertilization. Surrounding maternal tissues join with endosperm cells to form the protective seed coat. The endosperm divides and produces food storage materials. The zygote begins to divide and form the embryo. As the cells divide in the monocots, they push into the endosperm in a single file, forming a single cotyledon. Two rows of dividing cells enter the endosperm of a dicot plant. As the cotyledons continue to divide, they absorb the endosperm and grow into the space of the embryo sac. A group of cells at the opposite end divide and produce the embryo. The cells at the cotyledon junction form the epicotyl. At its tip are shoot apical meristem cells. Some of these cells differentiate into the first embryonic leaves. The region adjacent to the epicotyl is the hypocotyl. At its tip is the radicle, containing root meristem tissue (Figure 16.2). The radicle develops into the primary root.

Seed dispersal. Plants are stationary organisms. They have evolved a variety of mechanisms to ensure seed dispersal. Winged and plumed seeds are carried on the wind. Coconuts are an example of water-borne seeds. Animals can carry seeds in two ways. Sticky or barbed seeds adhere to animal coats. Fruits with fleshy interiors are eaten by animals; some seeds pass through the digestive tract untouched and are deposited with other animal wastes.

CHAPTER REVIEW

1. List the photosynthetic pigments, carbohydrate storage products, and structures found within the Subkingdoms
   a. Rhodophycophyta
   b. Phaeophycophyta
   c. Euchlorophyta

2. List the major characteristics of the four Euchlorophyta divisions.

3. List the major characteristics of the five subdivisions of the Trachaeophyta.

4. Differentiate between the gymnosperms and the angiosperms.

5. Differentiate between monocots and dicots.

6. Define mechanisms of water, mineral, and gas intake in terrestrial plants.

7. Define mechanisms of nutrient transport in terrestrial plants.

8. Describe the processes of seed dispersal and seed germination.

9. Describe primary and secondary growth of roots and shoots.

10. Describe the structures of leaves and flowers.

11. List five plant hormones and define the regulatory role of each.

12. Define asexual reproduction and give eight examples.

13. Outline the steps in the production of
    a. male pollen cells
    b. female egg cells

14. Describe the process of double fertilization and its products.

15. Describe fruit and seed development.

KINGDOM ANIMALIA

This chapter presents a summary, phylum by phylum, of the organisms within the Animal Kingdom. In some cases, classification to the "suborder" level is given. It is intended to serve as a survey of the organisms within the Kingdom Animalia. Specialized cells, organs, and functions are highlighted within each phylum. Classification schema within this kingdom are fluid — they are still argued over and changed. Therefore, the classification presented here does not agree with all texts. In those cases where a group of organisms is known by more than one name, several or both names are presented.

Classification in the Kingdom Animalia is largely based on embryology and body plan. This implies more than structure; it includes a variety of body functions as well.

All metazoa (animals) have three layers of cells in the early gastrula stage of development. These are the **ectoderm** — the outermost layer of cells, which develop into the outer covering and neural tissue; **endoderm** — the innermost cell layer, which develops into the gut, gut epithelium, and outpocketing of the gut; and **mesoderm** — the middle cell layer, which gives rise to skeletal, muscular, vascular, renal, and connective tissues.

Type of body cavity (coelom) is the basis for the primary subdivision of organisms within the kingdom. **Acoelomates** are animals without a coelom. **Pseudocoelomates** have cavities filled with fluids, but have no linings of mesodermal origin. **Coelomates** have a true coelom, lined by mesodermally derived peritoneum.

ACOELOMATES

Phylum Porifera

The **sponges** are radially symmetrical aquatic animals. Their body structure is quite simple: the vase-like body remains attached to the substratum (sessile) and has a large opening at the top. Water passes through pores, carrying with it oxygen and microscopic bits of food and organic matter.

The cells of sponges are not organized into tissues or organs, but include specialized cell types. **Choanocytes,** (collar cells) line the body cavity and set up water currents by flagellar motion. Epitheliomuscular cells, individual contractile cells, cover the outer surface. There is no coordination among these cells. The middle cell layer contains undifferentiated **archaeocytes**. Some archaeocytes form **amoebocytes**, cells that carry digested food from inner to outer surfaces; others form cells that secrete silicon or calcium carbonate spicules within the body; still others differentiate to produce reproductive cells. Sponges reproduce asexually by two means: formation of branches that break away from the parent; or production of **gemmules**, flagellated archaeocytes within a protective coat.

Phylum Coelenterata

The **Coelenterates**, or **Cnidarians** (corals, anemones, hydra, and jellyfish) possess radially symmetrical hollow bodies structured one of two ways: as the vase-like **polyp** (sessile) or the bowl-like **medusa** (motile). The outer body layer, **epidermis**, is separated from the inner **gastrodermis** by the gelatinous mesoglea.

Animals in this phylum possess a **coelenteron** (digestive cavity). Enzymes released into the coelenteron begin partial digestion of food. Gastrodermal cells phagocytose the smaller food particles and pass the nutrients to the rest of the body cells. Coelenterates are carnivorous, capturing and eating other living animals. Specialized **cnidoblast** cells, within the tentacles circling the coelenteron, help capture prey. Chemical or mechanical stimuli cause the cells to discharge **nematocysts** (thread capsules) that may be poisonous, sticky, or harpoonlike.

Coelenterates also exhibit rudimentary nervous systems and specialized receptor cells. Cnidoblasts are **independent effectors**, that is, they receive and respond to stimuli as individual cells. In hydra, a simple *nerve net* underlies the epitheliomuscular cells of the epithelium. These nerve cells synapse only with adjacent cells and can pass impulses in either direction (nonpolar). The medusa has a condensed nerve ring and two types of sense organs: **statocysts**, which help the animal to orient to gravity, and light sensitive **ocelli**, groups of pigment and photoreceptor cells.

Phylum Ctenophora

The luminescent **Ctenophores** (comb jellies and sea walnuts) are similar in body structure to coelenterates. Undulation of fused cilia (combs) pushes these animals along. Secretions from glue cells in the tentacles help capture plankton and larvae as food.

Phylum Platyhelminthes

The **Platyhelminthes** (flatworms) are *bilaterally symmetrical*, with definite anterior and posterior ends. Nerves are concentrated in the anterior end. Some species have fully developed organs and organ systems. Flatworms lack specialized respiratory and circulatory systems, however, and must rely instead on their flattened body structure for diffusion of oxygen. A pouch-like gastrovascular cavity, similar to the coelenteron, functions as an incomplete digestive system. Flatworms have specialized reproductive organs, gonads, and an excretory system. The three germ layers — ectoderm, mesoderm, and endoderm — are well defined. These animals also have muscle layers that are mesodermally derived, rather than epitheliomuscular cells.

The Platyhelminthes are further divided into three classes: **Class Turbellaria** — free-living planaria; **Class Trematoda** — parasitic flukes; and **Class Cestoda** — parasitic tapeworms.

Phylum Nemertinea

The **Nemertineans** are free-living marine predators. They have a complete digestive system — food goes in the anterior end and waste goes out the posterior end. Nemertineans capture prey in a unique way, using a retractable proboscis that lies inside a cavity near the mouth. The proboscis is forced out by fluid pressure, the prey is harpooned, and the proboscis is reeled back with the catch.

PSEUDOCOELOMATES

The **Pseudocoelomates** possess a complete digestive system with a one-way gut. Tissues surrounding the gut, however, are neither mesenchymal (as in the acoelomates) nor mesodermal (as in coelomates). Rather, the gut is surrounded by a fluid-filled cavity, the **pseudocoel**, a persistent embryonic blastocoel. Internal organs are free within the pseudocoel, which appears to function in circulation and the maintenance of turgor.

Phylum Aschelminthes

Most species of **Aschelminthes** (rotifers and nematodes) are elongate, cylindrical, and bilaterally symmetrical. The body is covered with a collagen cuticle. **Protonephridia**, which form a simple renal system, are often present.

Class Nematoda. The **nematodes** (roundworms), include parasitic members with a muscular pharynx and strong lips. They attach to the host's intestines and suck food, blood, and cellular fluids. Others are plant parasites. Most species, however, are free living, and found in every conceivable moist or aquatic habitat.

Nematodes are **dioecious** — the sexes are separate with distinct male and female individuals. They range in size from microscopic to 35 cm for the female Ascaris worm.

Class Rotifera. The **rotifers** are not parasitic, but **filter feeders**. Ciliary action sweeps microorganisms into their mouth. Their digestive system consists of specialized organs, stomach, glands, intestines, and anus. Rotifers reproduce both sexually and asexually. Fertilization is internal. Males are scarce, absent in some species. Females are able to reproduce parthenogenetically (without fertilization). Offspring are females only, and environmental conditions determine whether they will be of the asexual or sexual reproductive type.

Other Aschelminthes

Miscellaneous other pseudocoelomates are also classified as Aschelminthes. Some classification schema separate these organisms into their own phyla.

Phylum Priapulida

Priapulids are common in Cambrian fossil deposits. Today, there are only eight living species. Larval forms are indistinguishable from adult rotifers.

COELOMATES

Phyla Phoronida, Bryozoa, Entoprocta, and Brachiopoda

These phyla are sometimes classified together as the **Lophophorates**, because the lophophore (food-catching organ) is characteristically present in all members. A U-shaped or circular fold of the body wall encircles the mouth. Cilia on the lophophore tentacles create water currents that carry food and oxygen into the body cavity. Almost all species are sessile. The lophophores are well represented in Cambrian fossil deposits.

Phylum Phoronida

These wormlike animals live within a chitonous tube buried in the sand or attached to a substratum. They grow to less than 20 cm. Characteristic of sessile coelomates, the digestive tract is U-shaped. Contractions of the dorsal and ventral vessel move blood through the vascular system. Most species are hermaphroditic (both male and female sex organs are found in each individual).

Phylum Bryozoa

The **bryozoans** are colonial, sessile organisms. Because of their small size (usually less than 0.5 mm), they have no gas exchange, circulatory, or excretory systems. In many species, an **operculum** covers the anterior body opening. The operculum opens when the lophophore is extended and closes when the lophophore is withdrawn into the body. Some species possess a body covering composed of chiton, which overlays the calcium carbonate covering secreted by the epithelium. Most species are hermaphroditic and brood their eggs within the coelomic cavity or within external chambers.

Phylum Entoprocta

These species were once included within Phylum Bryozoa but are now usually placed within their own phylum. Some question exists as to whether these animals are pseudocoelomates or coelomates. All species are marine and live attached either to the ocean floor substratum or to other living organisms. They never exceed 5 mm in length. The **calyx**, an oval-shaped body structure containing the internal organs, is attached by one or many stalks. Both the mouth and the anal openings are located within the circular tentacles of the lophophore. Asexual reproduction by budding is common. The phylum contains both dioecious and hermaphroditic species.

Phylum Brachiopoda

The **brachiopods** (lampshells) resemble mollusks. However, unlike mollusks, these shelled lophophores hinge the body dorsoventrally (top to bottom) rather than laterally. The animal is usually attached to the substratum by means of a pedicle. All species are marine, few live beyond continental shelf depths. Most Brachiopod species are dioecious; eggs and sperm are shed into the water where fertilization takes place. A few species brood their eggs. Brachiopods are common Paleozoic and Mesozoic fossils.

Phylum Annelida

The **annelids** are the **segmented worms**.

Class Oligochaeta. The **oligochaetes** (earthworms) exhibit not only external segmentation but internal repetitions as well. Elements of nervous, circulatory, digestive, and excretory systems are found within each segment.

The nervous system consists of a *ventral nerve cord* with side branches and paired *ganglia* (masses of nerve cells). A large mass of nerve tissue, *brain*, in the anterior segments, is an example of **cephalization** (specialization of body to include a distinct head region).

Earthworms have a closed circulatory system — blood is enclosed within vessels, not merely pooled in coelomic spaces. Two large vessels run the length of the earthworm, one dorsally, the other ventrally. Irregular muscular contraction of five hearts, located in anterior segments, pumps the blood. Valves in the dorsal vessel and the hearts prevent backflow of blood. Hemoglobin is dissolved directly in the circulating fluid and is not associated with particular blood cells.

The digestive system has several specialized regions

for swallowing, grinding, and storage of food. The intestine functions in digestion and absorption. Earthworms feed on soil and are of primary importance in aerating and fertilizing soil.

The excretory system consists of ciliated funnels that open into the coelomic cavity and are connected to *nephridia* tubules within each each segment. Ciliary action sweeps the coelomic fluid into the tubule. Salts, sugars, and other nutrients are resorbed by the tubule walls. The remaining waste materials are passed through *pores* in each segment.

Gas exchange takes place on the moist body surface. No special respiratory organs are present.

Earthworms are hermaphroditic, though not self-fertilizing. A cylindrical band of tissue, the clitellum, holds the worms together during copulation. Two to three days after sperm are deposited and the eggs fertilized, the clitellum secretes a mucus sheath that surrounds the egg mass, protecting them until they are deposited.

Class Hirudinea. The **Hirudinea** (leeches) are primarily parasitic, however, some predatory (scavenger) species exist. They are **ectoparasites**, attaching to the outer surface of their host. Their coelom is not segmented although segmentation of their functional systems does exist.

Class Polychaeta. The polychaetes (many-haired or bristle worms) are common along seashores where they burrow into sand and mud. Most species are hermaphroditic. Fertilization is external. Eggs and sperm are shed into the water, and the zygote develops there into a ciliated larva. The Polychaetes are found in Precambrian fossil deposits.

Phylum Mollusca

Although Phylum Mollusca is subdivided into six classes, all members share certain characteristics. A muscular foot, present in all mollusks, is used for a variety of purposes — movement, digging, and capturing prey. All species have a fleshy mantle, which secretes an external covering (shell) in some species. Aquatic species possess respiratory gills inside the mantle cavity. In terrestrial species, gas is exchanged across the moist lining of the mantle cavity.

The coelomic cavity of adult mollusks is limited to an open cavity surrounding the heart. It does not extend, tubelike, through the body as in the Annelida. The circulatory system is open. Blood leaves the vessels and collects in spongelike sinuses before returning to the heart. Mollusks range in size from minute to over 18 m in length for some species of squid.

Four classes of Phylum Mollusca are discussed below.

Class Amphineura. The **amphineurans** (chitons) are the least specialized mollusks. A shell made of eight plates covers the flattened body. The foot is used for slow movement along intertidal rocks or to grip them as a defense mechanism. Gills fill the groove between the mantle cavity and the foot. A scraping radula removes surface algae and directs the food into the digestive tract.

Class Gastropoda. Both terrestrial and aquatic species of the **gastropods** (stomach-foot) exist. These include shelled forms (snails) as well as nonshelled forms (slugs — terrestrial; nudibranchs — aquatic). Their large muscular foot lies just below the digestive organs. During development from the larval stage, the animal undergoes a 180-degree rotation or torsion, which twists both the shell and organs therein, to create space for retraction of the head and foot into the shell. An operculum acts as a door, closing the organism inside it shell.

Class Pelecypoda. The **bivalves** are the two-shelled mollusks. Siphons suck water into the mantle cavity and over the gills. Labial palps near the mouth separate nutrients from non-nutrients. Food is channeled through a stomach and extensive intestine that passes through the heart before terminating at the anus. Powerful muscles hinge the lateral shells.

Class Cephalopoda. The squid and octopus belong to the **Class Cephalopoda** (head-foot). Most of these predators lack an external shell. Unlike other mollusks, they have a closed circulatory system. The muscular mantle serves to pump water over the gill surface. The molluscan foot has been modified into tentacles. They move in a jetlike fashion by forcing water through their siphons. Cephalopods have eyes similar in structure to vertebrate eyes. In addition, the octopus has a well-developed brain.

Phylum Onychophora

The **Onychophorans** share many traits with annelids and arthropods (yet to be discussed). They are considered by many to be an evolutionary link between the two phyla. Onychophorans have a smooth body wall, but internal and appendicular segmentation is evident. Their soft appendages terminate in claws. Breathing is done through a series of spiracles (holes) that open into tiny tracheal tubes. They have an open circulatory system. All living species are terrestrial, though many marine Onychophorans are found in earth Cambrian fossils.

Phylum Arthropoda

The **Arthropods** (jointed-leg) are the largest of the phyla in terms of sheer numbers. Their bodies are segmented and specialized for specific tasks. An **exoskeleton** (exo = outside) made of chiton and calcium salts covers the body. They have an open circulatory system. Arthropods obtain nutrients through a variety of means. They can be scavengers, ectoparasites, or endoparasites, omnivores, herbivores, and carnivores. They have established themselves in nearly every aquatic and terrestrial habitat on Earth. Their great success is due in part to their reproductive capacity and the genetic variation possible in sexual reproduction.

The Arthropods are divided into two subphyla based on the number of appendages and the mouth or feeding structures.

Subphylum Chelicerata. The **chelicerates** lack jaws. Instead, the first pair of pre-oral appendages have been specialized into feeding structures called chelicerae. The first pair of post-oral appendages, the pedipalps, have been modified to serve a variety of functions. Chelicerates do not have antennae. Their bodies are divided into a cephalothorax and abdomen.

Class Merostomata (horseshoe crab) are the aquatic chelicerates. Abdominal appendages have been modified into flaplike gills that are fused along the midline. The leaflike folds function as the site of gas exchange in these special book gills.

Class Pychnogonida or **Pantopoda** (sea spiders) are marine animals. Brain, sense organ structure, and presence of chelicerae place them within this subphylum. However, they also possess aberrant structures not found elsewhere within the Chelicerate subphylum. These include ovigerous legs (legs for egg brooding on the male), additional walking legs, and a segmented trunk.

Class Arachnida (spiders, mites, ticks, and scorpions) are terrestrial chelicerates. They have six pairs of appendages: one pair modified as chelicerae, one pair modified as pedipalps, and four pairs of legs. They are carnivorous: digestive enzymes are secreted into the prey and the resulting liquid nutrient broth is ingested. Gas exchange occurs across book lungs (modified book gills), trachae, or both. The book lung is an invagina-

tion of the ventral wall of the abdomen. Tracheal systems appear to be of two types: a sieve type derived from the book lung, or spiracles opening into an atrium or directly into a tracheal tube. The arachnid heart is segmented. Muscle layers contract the heart and dilation occurs as a result of the body's pull on ligaments suspending the heart within the coelom. Arachnids are dioecious. Sperm transfer is indirect; that is, a spermatophore structure containing sperm is placed within the soil and a female takes it up within her abdominal gonopore.

Subphylum Mandibulata. The arthropods of Subphylum Mandibulata have jaws or mandibles, antennae, and paired appendages, at least three of which are walking legs.

Class Crustacea includes both freshwater and marine organisms (shrimp, lobster, crabs, crayfish, copepods, barnacles, pillbugs, and wood lice). Crustaceans have two pairs of antennae, a distinguishing feature of this class. The rigid exoskeleton contains calcium salts, unlike the cuticle of other arthropods. Modification of appendages permits swimming, crawling, and burrowing. Many crustaceans are filter feeders; water currents are set up by modified appendages rather than cilia. Other forms of feeding include scavenging, hervibory, and carnivory. Many crustaceans rely on more than one form for obtaining nutrients. The circulatory system is similar to that of the chelicerates. Many species utilize accessory hearts to increase pressure. Gas exchange is carried out across gills, usually associated with the appendages. Crustacean sense organs include both media (inverse pigment-cup ocelli) and compound eyes (eyes containing a large number of **ommatidia** visual units), chemoreceptors and tactile receptors and **statocysts** (for maintenance of balance). Most crustaceans are dioecious, copulate, and brood their eggs.

Class Chilopoda (centipedes) are terrestrial predators. Modification of the first appendage results in a poisonous claw used for defense and in feeding.

Class Symphyla contains small (2-10 mm) terrestrial organisms living in soil and leaf mold. They superficially resemble the centipedes. They have a single anterior spiracle and a spinning organ, but have no eyes. Symphylans are dioecious. Sperm transfer is indirect (via spermatophores). Parthenogenesis is common.

Class Diploda (millipedes) has double trunk — segments, each bearing two pair of appendages. They vary in length from 2 mm to 28 cm. Repugnatorial glands (one per segment) aid in defense by exuding a variety of noxious substances — hydrogen cyanide, phenols, aldehydes, quinones, and a variety of other caustic substances. Millipedes are primarily herbivorous.

Class Pauropoda, similar to millipedes, are small soft-bodied grublike animals found in leaf mold and soil. They have no heart nor tracheal system due to their small size.

Class Insecta or **Hexapoda** is the largest class of animals. Distinguishing features include three pair of legs, two pair of wings (some parasitic species such as lice and fleas are wingless), a single pair of antennae, a pair of compound eyes, and a tracheal system for gas exchange. The body is divided into three segments: the *head* with downward oriented mouthparts, the *thorax* with the three leg appendages, and the *abdomen* with modified appendages and reproductive structures. The mouth parts are highly modified to accommodate many forms for obtaining nutrients — sucking, piercing, chewing, and cutting.

Accessory hearts are found in head, thorax, legs, and wings. The blood of insects is usually green or colorless. Organic molecules (rather than inorganic ions, as in most animals) function as osmotic regulators within the blood. The tracheal system terminates in **tracheoles** (1 micron in diameter) that branch out over tissue cells. Tracheole fluid is the final oxygen transport vehicle. Malpighian tubules lying free within the **hemocoel** (blood cavity) are the primary excretory organs. A waxy cuticle covering prevents excess water loss.

The insect brain is divided into three parts and numerous sense organs are distributed over the body segments. **Tympanal organs** (auditory receptors) are found in those species with sound-producing organs, as in crickets, grasshoppers, and cicadas.

Insects are dioecious. Sperm transfer may be direct or indirect, depending on the species. Development is of two types. **Gradual** or **incomplete metamorphosis** involves growth from the immature stages to the adult stage while maintaining similar body structure. (Immature terrestrial insects are called **nymphs**, immature aquatic, **naiads**.) **Complete metamorphosis** involves three stages. The larval stage is wingless. The pupa is the nonfeeding, quiescent stage in which the insect is usually hidden in a protective spot. During this time, adult structures and organs are formed. The adult form is winged. This process is under endocrine control.

A number of insect species exhibit social behavior, a type of functional interdependency.

Phylum Pogonophora

This is a relatively new phylum, unknown before 1900. The **pogonophorans** (beardworms), are sessile, deep water animals. They range in length from 10 to 85 cm. Their long tentacles bear minute extensions of individual epithelium cells, called pinnules. They lack a mouth and digestive system. Cilia lining the tentacles are thought to sweep nutrient-laden water through the body. Enzymes are secreted near the pinnules where extracellular digestion takes place. Pogonophorans are dioecious. Sperm transfer is indirect.

Phylum Chaetognatha

The **chaetognaths** (arrowworms), are common marine plankton animals. Their body is divided into a head region separated from the trunk by a narrow neck, and a tail region. Curved chitonous spines hang from the head and are used to capture prey. A fold in the body wall (hood) can be pulled forward to cover the head and protect the spines when they are not in use. The digestive tract is simple. No excretory, circulatory or gas exchange organs exist. Chaetognaths are hermaphroditic. Some species are self-fertilizing.

Phylum Echinodermata

The **echinoderms** (spiny-skinned animals), are bottom dwelling marine animals. Their body exhibits **pentamerous** (five part) **radial symmetry**. However, they are not related to the other radially symmetric phyla; rather, their symmetry is a modification of the bilateral body structure. Echinoderms possess an **endoskeleton** (*endo* = internal) of calcareous plates secreted by the dermis and covered by an epidermis. Distinctive features of the phylum are the water vascular system, a system of canals within the coelom, and the tube feet or podia that assist in a variety of functions — locomotion, excretion, and gas exchange. Most species are dioecious. Fertilization is external.

Class Stelleroidea; Subclass Asteroidea. The **Stelleroidea** contains free-moving, star-shaped animals whose arms project from a central disk. The **Asteroidea** (sea stars and starfish) possess five arms, although seven, fourteen, and forty arms are present in some species. Arm length in most species is one to three times the diameter of the central disc; width of the arm increases from tip toward the central disc. Asteroidea range in size from less than two centimeters to one meter across.

The mouth is found on the ventral oral surface of the body in the center of the disc. Furrows, containing tube feet, extend from the mouth to the tip of each arm. Asteroids are carniverous. The stomach occupies a central position and the pyloric caeca, composed of digestive glands, radiate into each arm.

The tube feet (and appendages of the body wall) play an important role in both excretion and gas exchange. Ammonia wastes are engulfed by coelomocytes that migrate to the distal tips of the papulae where they are ejected from the body. The ciliated peritoneal lining of the papulae produces a water current over the surface of the thin-walled papulae and the tube feet, allowing gas exchange to take place.

Asteroids have eye spots on the end of their arms. Epidermal sensory cells function as light, contact, and chemical receptors.

Asteroids exhibit **regeneration**, the ability to reform body parts or the entire body from an isolated part. They normally reproduce by **fissiparity**, an asexual mode of breaking in two with subsequent regeneration. A few species are **protandric hermaphrodites**, males when young that develop into females as they increase in size and age.

Class Stelleroidea; Subclass Ophiuroidea. Subclass **Ophiuroidea** (brittle stars and serpent stars) have extremely long arms, in contrast to the Asteroidea. Typically they have five arms that branch extensively in some species. The tube feet, one pair per arm, do not function in locomotion. Rather two lateral (side) arms rapidly move back and forth across the substratum while the other arms "lead" or "trail" the animal.

Ophiuroids are bottom dwellers, feeding by scavenging, browsing, or filter and deposit feeding. Little is known about their simple digestive tract. The stomach occupies the central disc; no intestine or anus is present, no part of the digestive tract extends into the arms.

Gas exchange occurs through thin-walled respiratory bursae lined with ciliated epithelium. These invaginations of the oral surface may also function in waste removal.

Ophiuroids possess no special sense organs.

Automization, breaking off of arms when disturbed or seized by a predator, is common among ophiuroids. They reproduce asexually by fissiparity. Many species are protandric hermaphrodites. Dioecious species exhibit no sexual dimorphism.

Class Echinoidea. Echinoideans (sea urchins, sea biscuits, heart urchins, and sand dollars) have circular-shaped bodies covered with spines, hence the name Echinoidea, meaning "like a hedgehog." Some species are radially symmetrical; others have attained secondary bilateral symmetry.

The echinoid water vascular system and nervous system are similar to those of the asteroids. Tube feet also function in locomotion.

Sea urchins possess a chewing apparatus known as **Aristotle's lantern**, composed of five calcareous plates projecting as teeth. Muscular control allows opening and closing of the teeth, as well as protraction and

retraction through the mouth opening. Sea biscuits, or heart urchins, lack such an organ; however, a modified structure exists in the sand dollar.

Coelomocytes within the principal coelom and coelomic subcompartments function in nutrient transportation and waste removal. Five pair of gills, cilia-lined epithelial outpockets of the body wall, function in gas exchange in the sea urchin. Thin-walled modified podia act as gas exchangers in the sea biscuit and sand dollar.

Echinoids are dioecious and most exhibit no sexual dimorphism.

Class Holothuroidea. Distinguishing characteristics of the **holothuroids** (sea cucumbers) include lack of arms, a polar body axis with mouth and anus at opposite poles, a reduced skeleton consisting of microscopic spines and podia modified as oral tentacles. Sea cucumbers are burrowing animals. They are deposit feeders, trapping food particles in sweeping tentacles. Tentacles are inserted into the mouth one at a time and sucked clean. Coelomocytes play a dual role in the digestive tract; they carry enzymes into the intestinal lumen (cavity) and then transport nutrient monomers out of the intestine.

Gas exchange is carried out through respiratory trees, a system of tubules through which water circulates. Ammonia wastes are also diffused through the respiratory tubes. Particulate (solid) waste is removed via coelomocytes.

Some holothuroid species have been reported to undergo seasonal **evisceration** (rupture of either the anterior or posterior end with subsequent expulsion of internal organs located at that end), followed by regeneration of lost parts. The center for holothuroid regeneration appears to be the cloaca, comparable to the central disc of the asteroids and ophiuroids. A few species are protandric hermaphrodites; most are dioecious. Some species have special brooding pockets.

Class Crinoidea. Crinoid species include both sessile (sea lilies) and motile (feather stars) forms. The pentamerous body crown is held aloft by an attached stalk that has all but disappeared except in the sea lilies. (Paleozoic fossil species had stalks up to 25 m in length.) In contrast to other echinoderms, the oral surface faces upward. Arms radiate from the crown and appear to be jointed. Pinnules, jointed appendages, line each side of the arm. Crinoids are suspension feeders, they trap food particles in mucus secretion of the podia that extend from the pinnules. The water vascular system thus serves a primitive food-capturing function. Details of digestion are unknown. Wastes are gathered by coelomocytes. Gas exchange occurs across any thin-walled part of the body. Crinoids have considerable regenerative capabilities. All crinoids are dioecious, show no sexual dimorphism and possess no distinct gonads. Cold-water species brood their eggs in sac-like invaginations of the arm.

Phylum Hemichordata

The **hemichordates** (acorn worms) are soft-bodied, burrowing, marine organisms. Their body is divided into three parts, the proboscis (the anterior acorn-shaped nose), the collar, and the trunk; all parts contain a coelomic cavity. The external surface is ciliated and gland cells secrete copious amounts of smelly mucus. Gill slits are located in the pharynx wall in the anterior portion of the trunk, behind the collar. The nervous system consists of sheets of nerve fibers lying beneath the epidermis. In the dorsal part of the collar, it forms a rolled up, hollow **neural tube**. The hemichordates have no special sense organs. They are dioecious. Fertilization is external. Development is similar to that in echinoderms.

Phylum Chordata

Characteristics of the chordate organisms are a **dorsal hollow nerve cord**, and the presence at some stage in their development of: *gill slits*, *notochord* (flexible dorsal rod), and a *postanal tail*.

The chordates are divided into three subphyla.

Subphylum Urochordata. **Urochordates** (tunicates or sea squirts) show all four chordate characteristics only during the short larval stage (one to two days). They are absent in the adult. Urochordates are filter feeders, using the remnants of *gill clefts* as strainers. They have an open circulatory system and a heart that pumps blood in one direction and then the other. Tunicates secrete cellulose as their body covering; this carbohydrate polymer is rare in animals.

Subphylum Cephalochordata. **Cephalochordates** (lancets) are fishlike marine animals with obvious pharyngeal gill slits. The notochord remains in the adult lancet and functions as an anchor for the **myotomes**, muscle segments. Beating cilia draw water into the complex mouth where food is separated for digestion while the water continues on its way, passing over the gills and exchanging oxygen and carbon dioxide.

Subphylum Vertebrata. In addition to the chordate characteristics, most vertebrates share the following traits: a *vertebral column*, the differentiation of the anterior end of the nervous system into a *brain* with special receptors, a closed circulatory system with *chambered heart*, the concentration of *hemoglobin within blood cells*, further specialization of gills and lungs for gas exchange, and the development of a *mesodermal kidney* that functions in excretion and regulation of osmotic balance.

Class Placodermi are *extinct*. They were jawed, armored fish that had paired pectoral and pelvic fins. The jaw appears to have been derived from the bony gill arches which support the gills.

Living vertebrates are subdivided into seven classes.

Class Agnatha are the **ostracoderms** (jawless lampreys and hagfish). They have simple median fins. They feed by a rasping tongue or device, aided in some species by a sucker mouth. Adult lampreys are parasitic whereas adult hagfish are bottom scavengers. The ostracoderms lack a bony skeletal system. Lack of bone in these animals is thought to represent a secondary loss.

Class Chondrichthyes are the **cartilaginous fish** (sharks, rays, skates, and chimeras). They are largely predators and scavengers. Placoid scales, miniature teeth, are embedded in the skin of sharks and rays. The teeth of the mouth and the placoid scales grow, migrate, and fall out when old. Many sharks are filter feeders. The cloaca is the common terminus for both the urinary and reproductive systems. The vent is the cloacal opening to the exterior. As all fish, sharks have a **lateral line organ** sensitive to the low frequency sound waves characteristic of water movements. Other sensory cells respond to electrical fields created by nearby animals. Sharks lack the muscles for propelling water past the gills. Therefore, they must keep moving to pass water into the mouth and over the gills. Fertilization is internal.

Class Osteichthyes are the **bony fish**. Their fin musculature is more extensive allowing greater maneuverability. Bony fish also have an **air bladder** that functions in balance and buoyancy. Most fish produce scales as their outer covering. Sensory organs include the lateral line organ, eyes, taste buds, and balance centers. The gills are located within a gill chamber protected by a movable external bony operculum. The bony fish have a two-chambered heart. Fertilization in most species is external.

Class Amphibia (salamanders, frogs, toads, and wormlike caecilians) reproduce and develop in wet or aquatic environments. An important source of gas exchange is moist vascularized skin. Some salamanders totally lack lungs. Amphibians have a three-chambered heart, and pump blood to the lungs for oxygenation before returning it to the heart for repumping to the entire body. Fertilization is external in frogs and toads, internal in salamanders and caecilians.

Class Reptilia (snakes, lizards, alligators, crocodiles, and turtles) are not bound to the water for any stage of their life cycle. Specialization within this class include development of: a penis for internal fertilization, a porous leathery shell for the egg (protecting it from water loss), food storage within the egg (yolk), skin covered by dry scales that reduces water loss in the adult, and well-developed lungs. Most reptiles are carnivorous. Reptile species ruled the Earth during the Mesozoic Era, the "Age of the Dinosaur."

Class Aves (birds) possess modifications permitting flight: specialization of scales to form feathers, hollow bones resulting in a light weight skeleton, and an enlarged breastbone that serves as a keel for the attachment of flight muscles.

Birds have a four-chambered heart, more efficient for separating unoxygenated blood from oxygenated blood. Large yolky eggs are produced protected by a hardened (not leathery) shell. Birds exhibit a great deal of parental care for their young.

Class Mammalia are characterized as the hairy animals that produce milk. They maintain constant body temperature. Fertilization is internal as is early growth and development. A muscular diaphragm moves air in and out of the lungs. The heart, as in birds, has four chambers. Limb structure has changed; limbs are directly under the body. Milk produced by modified sweat glands is a high-protein, high-caloric food for the young. Parental care for the young is extensive.

CHAPTER REVIEW

1. Describe the basis for metazoan classification.
2. Define the three types of metazoan body cavities.
3. Describe specialized cells, tissues, and sexual reproduction in the following groups:
   a. Phylum Porifera
   b. Phylum Coelenterata
   c. Phylum Platyhelminthes
   d. Class Nematoda
   e. the Lophophorates
   f. Phylum Annelida
   g. Phylum Mollusca
   h. Class Chelicerata
   i. Class Crustacea
   j. Class Insecta
   k. Phylum Echinodermata
4. Give the common name of a member of the above groups.
5. Compare and contrast the following features for each of the six Molluscan classes:
   a. body structure
   b. "foot" structure
   c. respiration
   d. circulation
6. Describe the different body structures of the Arthropods.
7. Compare and contrast the members of the Phylum Echinodermata. Include characteristics such as:
   a. body shape
   b. endoskeleton
   c. tube feet
8. How are the Hemichordates similar to the Chordates?
9. List the four characteristics of the Chordates.
10. List major characteristics of the three chordate subdivisions.
11. List the major characteristics of the Class Vertebrata.
12. Describe the eight classes of vertebrates.

Ecology is the study of the interactions between organisms and their environment. An **environment** is a multidimensional system encompassing all the physical and biological factors that either affect the organisms inhabiting it or are influenced by them. The most fundamental unit of study in ecology is the **population**, an interacting group of organisms of the same species that inhabit the same place at the same time (Chapter 12). That place where a population (or organism) lives is its **habitat**. A collection of all populations inhabiting the same place at the same time is a **community**. An **ecological system** or **ecosystem** includes all *biotic* (living) and *abiotic* (nonliving) components of a community and its physical environment. Finally, the **biosphere** is the total sum of the ecosystems of this planet.

POPULATION DYNAMICS

Size and Distribution

The number of individuals of a population in a given area or volume of space is termed the **crude density** of the population. However, when density is defined in terms of the number of individuals occupying the habitat specific for that species, it is termed **ecological density**. Because the distribution of individuals changes in time and space, ecological density can vary even though crude density remains constant.

Within their habitat, the individuals of a population may be distributed randomly, uniformly, or in clumps. The presence of uniform environmental conditions and the absence of interactions among individuals within a population is thought to contribute to the random distribution of individuals within a habitat (spiders

on a forest floor, for example). In contrast, competition among individuals of a population can result in their uniform distribution throughout the habitat, providing uniform environmental conditions prevail (creosote bushes in the scrub desert, for example). Clumping is the most common distribution for both plants and animals. It is a nonrandom response to interactions between organisms of different species having overlapping habitats, as in predation. Clumping can also result from reproductive and developmental mechanisms that cause offspring to be established close to members of their own species.

Population distribution also varies with time. For instance, seasonal variations in environmental conditions result in fluctuations in food availability for many habitats. Consequently, whole populations annually migrate to more productive areas.

Growth Parameters

Alterations in population size (ΔN) are described with the following formula:

$$\Delta N = (\text{natality} + \text{immigration}) -$$
$$(\text{mortality} + \text{emigration})$$

where **natality** = birthrate
immigration = number of individuals having joined the population
mortality = deathrate
emigration = number of individuals having left the population

In the absence of environmental constraints, all the individuals in each generation would achieve their full reproductive potential. As a consequence, population size would increase at a rate that is representative of **exponential growth** (Figure 18.1). Exponential growth usually occurs when a small population has access to abundant resources (for example, laboratory cultivated bacteria or protists).

Growth Limiting Factors. Factors that limit the size of a population are called **growth limiting factors** and are categorized into two types: density-independent and density-dependent. **Density-independent factors** are not influenced by population density (for example, rainfall, day length, or annual temperature extremes). **Density-dependent factors** are influenced by population density, and consequently have an overall homeostatic effect on population growth. When population size increases, density-dependent factors decrease natality and increase mortality; when population size decreases, they permit birthrate to increase and deathrate to decrease.

Density-dependent growth limiting factors are both abiotic and biotic. Some examples of abiotic factors are soil moisture, sunlight, and available nutrients. Biotic density-dependent factors involve both *intraspecific* (between members of the same population) interactions and *interspecific* (between members of different populations) interactions. Three interspecific (symbiotic) interactions that limit population growth are *competition*, *predation*, and *parasitism*. (Interspecific interactions are discussed later in this chapter.)

Carrying Capacity. The number of individuals in a population that can be supported by the environment for an indefinite period of time is termed the **carrying capacity (K)** of the environment. It is determined by the **environmental resistance** — all factors that limit population growth collectively. Figure 18.2 illustrates that a population grows exponentially until it reaches the carrying capacity of its habitat. For most populations, the

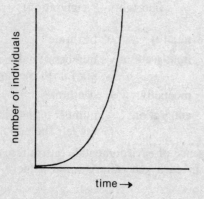

Figure 18.1. An exponential growth curve characteristic of small populations with access to abundant resources.

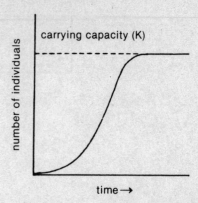

Figure 18.2. A logistic growth curve. The population grows exponentially at first, but as the size of the population approaches the carrying capacity of its environment, growth declines to a point that is in equilibrium with carrying capacity.

growth curves flatten out as carrying capacity is reached and a negative feedback relationship develops between population size and its rate of increase. This pattern of population growth is termed **logistic growth**.

If the exponential growth of a population is so dramatic that it greatly overshoots the carrying capacity of its habitat, the population will crash. A drastic reduction in population size will also result if the carrying capacity of the habitat is suddenly and dramatically reduced.

Regulation of Population Growth. Generally, there are two reproductive responses to growth limiting factors: r-selection and K-selection. Populations for whom density-independent factors are most important in limiting growth are termed **r-selected** (or **r-strategists**). They tend to inhabit variable, unpredictable, disturbed, or transitional environments, and are often subject to sudden and catastrophic mortality. Consequently, their fitness depends on high reproductive rates (numerous offspring) and rapid development of the young. They are generally small organisms that reproduce only once during short life spans. Because of these and related characteristics, r-selected populations are adapted to rapid dispersal and colonization — to new or changing environments.

For **K-selected** populations (**K-strategists**), density-dependent factors are most important in limiting growth. They tend to inhabit relatively stable and predictable environments. They develop more slowly and have longer lifespans than r-selected populations. K-selected populations reproduce more than once with fewer offspring produced and with a greater investment of parental care, in contrast to r-selected populations. Consequently, they tend to have low mortality in early life. They are not well-adapted for dispersal and colonization but tend to be larger in size and have greater competitive ability.

THE COMMUNITY: ORGANIZATION AND DYNAMICS

Ecological Niche

The various roles and associations of a species in the community of which it is a part are collectively described as its **ecological niche**. The concept is a multidimensional description of how a species interacts with the abiotic and biotic components of the ecosystem in which it lives and reproduces. The biotic components of a niche include the various interspecific interactions in which members of the species are involved.

Interspecific Interactions

Symbiosis describes all recurring interspecific interactions between two (or more) species. The effects of the interactions can be positive or neutral for one or all species involved. Interactions also differ as to degree of affiliation and as to how exclusive the relationship is. Seven forms of symbiotic interaction have been described: neutral, protocooperation, mutualism, commensalism, predation, parasitism, and interspecific competition.

Neutral interactions have no direct effect on the species involved. In **protocooperation**, the interaction is beneficial to all species involved. However, the relationship is *nonobligatory* (nonessential). For instance, honey bees and apple blossoms both benefit from the act of pollination but neither is dependent upon the other. **Mutualism** is similar to protocooperation in that all the species involved benefit; however, it is *obligatory* — essential to the survival of the species involved. For example, a lichen is a mutualistic association of fungus and green alga (or cyanobacteria). The fungus provides a hospitable environment for the photosynthetic species, which in turn provides the fungus with carbohydrates. Neither species can live without the other.

Commensalism is a symbiotic interaction beneficial for one species but neutral for the other. For instance, the use of trees for nesting benefits birds but does nothing for their hosts.

In **predation**, a **prey** species serves as the food source for a **predator** species. Obviously, the predator benefits and the prey is harmed. Some ecologists consider **parasitism** to be a form of predation because one species, the **parasite**, benefits, while another, the **host**, is harmed. Unlike predators, however, true parasites obtain nourishment from their hosts without killing them outright.

Finally, **interspecific competition** is the interaction that results from members of two (or more) species attempting to acquire the same limited resource. Research indicates that the more niches overlap, the more intense will be the competition, to the extent that no two species can occupy the same niche at the same time indefinitely (the principle of **competitive exclusion**).

THE ECOSYSTEM: STRUCTURE AND ENERGETICS

Trophic Levels

The various ways in which organisms acquire energy contribute, in part, to how ecosystems are organized. A **trophic level** of an ecosystem contains species that obtain energy primarily from a common source (Table 18.1). Energy moves through an ecosystem from one trophic level to the next. Energy is initially trapped by producers and passed on, unidirectionally, through a series of consumers, the sequence of which is termed a **food chain**. Each ecosystem is composed of a network of interconnecting food chains forming a **food web** (Figure 18.3).

Energy leaves ecosystems (primarily in the form of low grade heat) as a result of the metabolic activities of each producer, consumer, and decomposer. Consequently, some energy is lost at every trophic level. In fact, only about 10 percent of the energy at one trophic level is passed on to another.

Biogeochemical Cycles

Unlike energy, inorganic substances (e.g., water, carbon, nitrogen, and phosphorus) move through ecosystems as components of **biogeochemical cycles**. Inorganic substances are assimilated by primary producers from air, water, or soil, passed on to consumers, and eventually transferred to decomposers. Decomposers then release inorganic substances to the environment in forms that can be reassimilated by producers.

As nondegradable substances (e.g., DDT, mercury, and cadmium) move up through trophic levels, they tend to become concentrated. **Biological concentration** presents an increasingly serious hazard worldwide, as a consequence of environmental pollution. Toxins (insecticides, heavy metals, and radioactive substances, for example), which do not readily break down, enter ecosystems through food chains and become concentrated

Trophic level	Energy sources	Examples
Producers		
photosynthesizers	solar energy	cyanobacteria, protists, and plants
chemosynthesizers	oxidation of inorganic substances	nitrifying bacteria
Consumers		
primary consumers	producers	herbivores (planteaters)
secondary consumers	primary consumers	carnivores (flesheaters)
tertiary consumers	secondary consumers	high order carnivores (e.g., killer whales)
omnivores	producers, consumers, and decomposers	humans
Decomposers	organic remains and wastes of all other organisms	some bacteria, fungi, and invertebrates

Table 18.1. Trophic structure of ecosystems.

until they reach levels lethal to organisms of the highest trophic levels, including humans.

ECOLOGICAL SUCCESSION

The sequence of changes in species composition and trophic structure that an ecosystem undergoes is termed **ecological succession**. It involves a continuum of one community replacing another until the establishment of a **climax community** — a relatively stable, self-maintaining community. **Primary succession** begins in an area devoid of life, as in a newly formed volcanic island. **Secondary succession** occurs in a disturbed area previously inhabited, as in that devastated by fire.

As ecosystems mature, community biomass, species diversity, and the capacity for nutrient entrapment and retention increase, while productivity decreases. The capacity of an ecosystem to recover from disturbance also increases as it matures.

THE BIOSPHERE

The biosphere has three primary environmental components:

lithosphere — the outer portion of the Earth (60 to 100 kilometers in depth) that is divided into rigid plates

hydrosphere — the liquid and frozen water on or near the surface of the lithosphere (oceans, lakes, rivers, polar ice caps, etc.)

atmosphere — the region of gas, particulate matter, and water vapor enveloping the lithosphere and hydrosphere (50 kilometers in thickness)

Ecosystems are divided into two broad categories: *aquatic ecosystems* and *terrestrial ecosystems*. (No atmospheric ecosystems are known to exist.)

Aquatic Ecosystems

Aquatic ecosystems exist in the fluid components of the hydrosphere, and are numerous and diverse. Underwater topography, water depth, temperature, chemical content, and the degree and direction of water movement, all contribute to the diversity of aquatic ecosystems. There are three main types of aquatic ecosystems (based on salinity):

freshwater — those associated with inland bodies of standing freshwater (**lentic ecosystems**; e.g., lakes and ponds), and running water (**lotic ecosystems**; e.g., rivers and streams)

marine — those associated with coastal (**neritic**) and open (**oceanic**) regions of oceans

estuarine — those associated with coastal regions where freshwater from rivers and streams and seawater mix

Terrestrial Ecosystems

Terrestrial ecosystems exist in association with environments in the lithosphere. Differences in climate, topography, and abiotic composition result in corresponding differences in community structure. **Biomes** represent large collections of climax communities having common patterns of vegetation and climate. The major terrestrial biomes include various types of deserts, grasslands, shrublands, forests, and tundra.

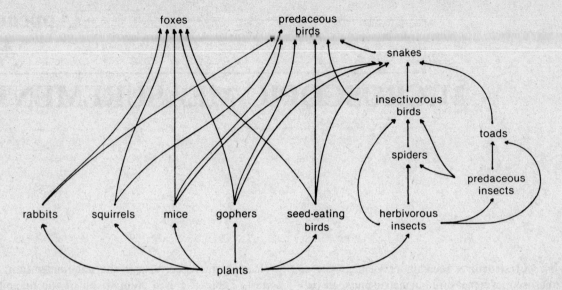

Figure 18.3.   Diagram of a simplified food web. In reality, many more organisms would be involved, including a host of decomposers.

CHAPTER REVIEW

1. Define the following terms:
   ecology
   population
   habitat
   community
   ecosystem
   biosphere
   growth limiting factors
   carrying capacity
   environmental resistance
   ecological niche
   biochemical cycles
   biological concentration
   trophic level
   food chain
   food web
   ecological succession
   climax community
   biomes

2. Contrast the following terms:
   a. crude density and ecological density
   b. density-independent growth limiting factors and density-dependent growth limiting factors
   c. exponential growth and logistic growth
   d. r-selected populations and K-selected populations
   e. predator and prey
   f. parasite and host
   g. primary succession and secondary succession

3. List the four major factors that influence the size of a population, define each, and explain how they are interrelated.

4. What is symbiosis? List the different forms of symbiosis, describing each.

5. Explain the competitive exclusion principle.

6. Outline the general trophic structure of ecosystems (beginning with the level at which energy enters the ecosystem). Describe each trophic level and its sublevels.

7. List and describe the primary environmental components of the biosphere.

8. Describe the two broad categories of ecosystems.

9. Contrast the three main types of aquatic ecosystem.

10. What factors determine the structure of terrestrial ecological communities? List some major categories of terrestrial biomes.

MICROSCOPIC MEASUREMENTS

The size of structures seen under visible light or electron microscopes is measured in metric units. In visible light microscopy, the **micrometer** or **micron** (μm; equivalent to 1/25,000 inch) is most frequently used; in electron microscopy, it is the **nanometer** (nm; 1/1000 μm). Table A.1 is a comparison of the magnitude of units of measurement used in visible light and electron microscopy.

Table A.1. Comparison of units of measurement used in visible light and electron microscopy.

	Millimeter (mm)	Micrometer (μm)	Nanometer (nm)	Angstrom (Å)
mm	1	1000	1,000,000	10,000,000
μm	0.001	1	1000	10,000
nm	0.000001	0.001	1	10
Å	0.0000001	0.0001	0.1	1

SOLUTIONS TO GENETICS PROBLEMS

1.

Genotypic ratio	Phenotypic ratio
a. all Tt	all tall
b. ¼ TT : ½ Tt : ¼ tt	¾ tall : ¼ dwarf
c. ½ TT : ½ Tt	all tall
d. ½ Tt : ½ tt	½ tall : ½ dwarf

2.

Genotypic ratio	Phenotypic ratio
a. all PpTt	all purple-flowered, tall
b. all PpTt	all purple-flowered, tall
c. ¼ PPTt	
¼ PpTt	
¼ PpTT	all purple-flowered, tall
¼ PPTT	
d. ¹⁄₁₆ pptt	¹⁄₁₆ white-flowered, dwarf
⅛ Pptt	
¹⁄₁₆ PPtt	³⁄₁₆ purple-flowered, dwarf
⅛ ppTt	
¹⁄₁₆ ppTT	³⁄₁₆ white-flowered, tall
¼ PpTt	
⅛ PPTt	
⅛ PpTT	⁹⁄₁₆ purple-flowered, tall
¹⁄₁₆ PPTT	

3. a. Because normal hands is a recessive trait (relative to syndactyly) and is only expressed in the homozygous state, both parents must have contributed a recessive allele to the offspring. Therefore, the mother must be heterozygous for syndactyly.

b. The probability that their first child would have normal hands is 50%. Since the birth of a second child is an independent event from the first birth, the probability that the second child would also have normal hands is again 50%.

c. The alleles for syndactyly are not sex-linked, and therefore assort independently of those on the sex chromosomes. The probability that the next child would be a girl with syndactyly would be the *product* of the individual probabilities for each trait (Second Law of Probability):

$$P_{(SsXX)} = P_{(Ss)} \times P_{(XX)}$$
$$= 0.5 \times 0.5$$
$$= 0.25 \text{ (or 25\%)}$$

d. In this question, the next two births are considered as a combined event, therefore the Second Law of Probability would apply:

$$P_{(ss, ss)} = P_{(ss)} \times P_{(ss)}$$
$$= 0.5 \times 0.5$$
$$= 0.25 \text{ (or 25\%)}$$

4. The ABO blood system is an example of codominance and multiple alleles. The O-type allele is recessive to the A- and B-type alleles, which are codominant. Thus, O-type blood is a homozygous recessive situation that necessitates both parents contributing an O-allele. Consequently, the father could be heterozygous for A- and B-type blood, because both would result in a 50% chance of contributing the O-type allele. However, the father could *not* have AB-type blood, because that would result in 0% probability of contributing the O-type allele.

5. a. Color blindness is a sex-linked recessive trait. In order for a female child to be born with the defect, the father must be color blind; for a male child, however, the father can have normal vision. In both cases, though, the mother must be at least a carrier of the trait ($X^G X^g$).

   b. Again, each birth is an independent event. The probability that the next child will be a normal-visioned male is 50%.

   parents: $\begin{matrix} G & g \\ X & X \end{matrix}$ X $\begin{matrix} g \\ X & Y \end{matrix}$

   color blind   normal-visioned

6. The genotype of the child would be $X^G X^g iiss$. The probability of each trait is as follows:

   normal-visioned girl $P_{(X^G X^g)} = 0.25$ (or 25%)

   O-type blood $P_{(ii)} = 0.5$ (or 50%)

   normal hands $P_{(ss)} = 0.5$ (or 50%)

   According to the Second Law of Probability, the probability that the next child would be born with the above genotype is:

   $$P_{(X^G X^g iiss)} = P_{(X^G X^g)} \times P_{(ii)} \times P_{(xx)}$$
   $$= 0.25 \times 0.5 \times 0.5$$
   $$= 0.0625 \text{ (or 6.25\%)}$$

REFERENCES

The following are suggested readings and sources of reference.

Avers, C. 1980. *Genetics*. rev. ed. Willard Grant Press, Boston.

Baker, J., and G. Allen. 1979. *A Course in Biology*. 3rd ed. Addison-Wesley, Menlo Park, California.

Barnes, R. 1980. *Invertebrate Zoology*. 4th ed. W. B. Saunders, Philadelphia.

Begon, M., and M. Mortimer. 1981. *Population Ecology: A Unified Study of Plants and Animals*. Sinauer Associates, Sunderland, Massachusetts.

Brock, T. 1979. *Biology of Microorganisms*. 3rd ed. Prentice-Hall, Englewood Cliffs.

Camp, P., and K. Arms. 1984. *Exploring Biology*. 2nd ed. Saunders College/Holt, Rinehart and Winston, New York.

Curtis, H., and N. Barnes. 1985. *Invitation to Biology*. 4th ed. Worth, New York.

Dickerson, R. 1978. "Chemical Evolution and the Origin of Life." *Scientific American* 239:70–86.

Dott, R., and R. Batten. 1981. *Evolution of the Earth*. 3rd ed. McGraw-Hill, New York.

Easton, T., and C. Rischer. 1984. *Bioscope*. 2nd ed. Charles E. Merril, Columbus, Ohio.

Farnsworth, M. 1978. *Genetics*. Harper & Row, New York.

Fox, S., and K. Dose. 1977. "Molecular Evolution and the Origin of Life." *Scientific American* 239:70–86.

Giese, A. 1979. *Cell Physiology*. 5th ed. W. B. Saunders, Philadelphia.

Hickman, C., et al. 1979. *Integrated Principles of Zoology*. 6th ed. Mosby, St. Louis.

Hopkins, R. 1981. "Deoxyribonucleic Acid Structure: A New Model." *Science* 211:289–291.

Keeton, W., and C. McFadden. 1983. *Elements of Biological Science*. 3rd ed. W. W. Norton, New York.

Kornberg, A. 1980. *DNA Replication*. W. H. Freeman, San Francisco.

Lake, J. 1981. "The Ribosome." *Scientific American* 245:84–97.

Lehninger, A. 1982. *Principles of Biochemistry*. Worth, New York.

Lodish, H., and J. Rothman. 1979. "The Assembly of Cell Membranes." *Scientific American* 240:48–63.

Mader, S. 1982. *Inquiry into Life*. 3rd ed. Wm. C. Brown, Dubuque, Iowa.

———. 1985. *Biology: Evolution, Diversity and the Environment*. Wm. C. Brown, Dubuque, Iowa.

Margulis, L. 1970. *Origins of Eukaryotic Cells*. Yale University Press, New Haven, Connecticut.

———. 1982. *Early Life*. Science Books International, Boston.

Mader, S., and K. Schwartz. 1982. *Five Kingdoms: An Illustrated Guide to the Phyla of Life on Earth*. W. H. Freeman, San Francisco.

Masterton, W., and E. Slowinski. 1977. *Chemical Principles*. 4th ed. W. B. Saunders, Philadelphia.

Moran, J., M. Morgan, and J. Wiersma. 1980. *An Introduction to Environmental Sciences*. 2nd ed. W. H. Freeman, San Francisco.

Miller, G. 1978. *Chemistry: A Basic Introduction*. John Wiley & Sons, New York.

Norstog, K., and A. Meyeriecks. 1985. *Biology*. 2nd ed. Charles E. Merrill, Columbus, Ohio.

Pianka, E. 1983. *Population Ecology*. 3rd ed. Harper & Row, New York.

Porter, K., and J. Tucker. 1981. "The Ground Substance of the Living Cell." *Scientific American* 244:57–67.

Raven, P., R. Evert, and H. Curtis. 1981. *Biology of Plants*. 3rd ed. Worth, New York.

Ricklefs, R. 1983. *The Economy of Nature: A Textbook in Basic Ecology*. 2nd ed. Chiron Press, New York.

Ritchie, D., and R. Carola. 1983. *Biology*. 2nd ed. Addison-Wesley, Menlo Park, California.

Schopf, J. 1978. "The Evolution of the Earliest Cells." *Scientific American* 239:111–138.

Sheeler, P., and D. Bianchi. 1980. *Cell Biology: Structure, Biochemistry, and Function*. John Wiley & Sons, New York.

Sherman, I., and V. Sherman. 1983. *Biology: A Human Approach*. 3rd ed. Oxford University Press, New York.

Singer, S., and G. Nicolson. 1972. "The Fluid Mosaic Model of the Structure of Cell Membranes." *Science* 175:720–731.

Smith, R. 1984. *Ecology and Field Biology*. 4th ed. Harper & Row, New York.

Stansfield, W. 1983. *Genetics*. 2nd ed. McGraw-Hill, New York.

Steffins, G. 1982. *Darwin to DNA, Molecules to Humanity*. W. H. Freeman, San Francisco.

Starr, C., and R. Taggart. 1984. *Biology: The Unity and Diversity of Life*. 3rd ed. Wadsworth, Belmont, California.

Stryer, L. 1981. *Biochemistry*. 2nd ed. W. H. Freeman, San Francisco.

Tortora, G., and N. Anagnostakos. 1984. *Principles of Anatomy and Physiology*. 4th ed. Harper & Row, New York.

Vander, A., J. Sherman, and D. Luciano. 1980. *Human Physiology: The Mechanisms of Body Functions*. 3rd ed. McGraw-Hill, New York.

Villee, C., E. Solomon, and P. Davis. 1985. *Biology*. Saunders College/Holt, Rinehart and Winston, New York.

Wallace, R. 1981. *Biology: The World of Life*. 3rd ed. Goodyear, Santa Monica, California.

Wallace, R., J. King, and G. Sanders. 1984. *Biosphere: The Realm of Life*. Scott, Foresman and Company, Glenview, Illinois.

Ward, J., and H. Hetzel. 1984. *Biology: Today and Tomorrow*. 2nd ed. West, San Francisco.

Weier, T., C. Stocking, and M. Barbour. 1982. *Botany: An Introduction to Plant Biology*. 6th ed. Wiley, New York.

Weinreb, E. 1984. *Anatomy and Physiology*. Addison-Wesley, Menlo Park, California.

Wheater, P., H. Burkitt, and V. Daniels. 1979. *Functional Histology: A Text and Colour Atlas*. Churchill Livingstone, New York.

Waese, C. 1981. "Archaebacteria." *Scientific American* 244:98–125.

Wolfe, S. 1981. *Biology of the Cell*. 2nd ed. Wadsworth, Belmont, California.

———. 1983. *Biology: The Foundations*. 2nd ed. Wadsworth, Belmont, California.

INDEX

ABOUT THE AUTHORS

Charles R. Wert received his B.S. in Biological Sciences from California Polytechnic State University at San Luis Obispo in 1974, with a concentration in Zoology. He also received his M.S. in Biological Sciences from Cal Poly in 1977. His graduate research involved an examination of the ability of juvenile red abalone (*Haliotis rufescens*) to acclimate to changes in water temperature. Mr. Wert received an M.A. in Education from San Diego State University in 1982, with a concentration in Community College Curriculum. He has been teaching in community colleges since 1982 and presently teaches general biology and oceanography at San Diego Mesa College, where he also serves on the Faculty Computing Committee. He resides with his wife, Margaret, in Lemon Grove, California.

Patty Kreikemeier-Gaffney received her B.S. degree in Zoology from Colorado State University, Fort Collins, Colorado. She graduated magna cum laude, in 1975. Her graduate work at the University of Cincinnati in the field of mammalian DNA replication led to an M.S. in Developmental Biology in 1979. Upon completion of her studies, she worked at the Salk Institute for Biological Studies in La Jolla, California, in the fields of tumor virology and regulatory biology. Ms. Kreikemeier-Gaffney earned teaching credentials in Life Science and Chemistry from San Diego State University in 1983. Currently, she teaches biology courses for the Grossmont Union High School District in the San Diego area.

Ms. Kreikemeier-Gaffney is married to senior systems engineer Thomas J. Gaffney, Jr. They are the parents of two daughters, Davida (3 years) and Mercedes (9 months). Together they enjoy singing, telling stories, and spontaneous picnic outings.